CHEMICAL VAPOR DEPOSITION FOR NANOTECHNOLOGY

Edited by **Pietro Mandracci**

Chemical Vapor Deposition for Nanotechnology
http://dx.doi.org/10.5772/intechopen.73342
Edited by Pietro Mandracci

Contributors

David Muñoz-Rojas, Viet Huong Nguyen, César Masse De La Huerta, Carmen Jiménez, Daniel Bellet, Chen Wang, Kizhanipuram Vinodgopal, GuiPing Dai, Marta Janus, Ausrine Bartasyte, Vincent Astié, Jean-Manuel Decams, Cyril Millon, Phuong Viet Pham, Tatsuhiko Aizawa, Yoshiyuki Suda

First published in London, United Kingdom, 2019 by IntechOpen
IntechOpen is the global imprint of INTECHOPEN LIMITED, registered in England and Wales, registration number: 11086078, The Shard, 25th floor, 32 London Bridge Street
London, SE19SG – United Kingdom
Printed in Croatia

British Library Cataloguing-in-Publication Data
A catalogue record for this book is available from the British Library

Additional hard copies can be obtained from orders@intechopen.com

Chemical Vapor Deposition for Nanotechnology, Edited by Pietro Mandracci
p. cm.
Print ISBN 978-1-78984-960-8
Online ISBN 978-1-78984-961-5

For EU product safety concerns:
IN TECH d.o.o., Prolaz Marije Krucifikse Kozulić 3, 51000 Rijeka, Croatia,
info@intechopen.com or visit our website at intechopen.com.

Meet the editor

Pietro Mandracci was born in Torino in 1970. He attained an MD in Physics at the Torino University in 1996 and a PhD in Electron Devices at the Trento University in 2001. From 2017 he has been an associate professor at Politecnico di Torino. During his career he has worked on the simulation of solar cells, on the fabrication of UV detectors based on amorphous semiconductors, on the design and fabrication of multilayer structures for photonic devices, as well as on the synthesis of advanced thin-film materials and nanostructures. His present research interests deal with thin-film technology and nanotechnology, structural, optical, and electrical characterization of materials and nanostructures, as well as their application to electronic and photonic devices, micromechanical systems, and biomedicine.

Contents

Preface

Chemical vapor deposition (CVD) techniques have played a major role in the development of modern technology since the start of the microelectronics industry, thanks to their ability to synthesize materials of high purity and high quality. In the last decade the rise of nanotechnology has further improved the importance of CVD technology, thanks to the ability of techniques such as atomic layer deposition (ALD) and vapor liquid solid growth to achieve the synthesis of nanodimensional structures.

This book presents some of the most interesting developments in CVD technology, considering both the fundamentals of novel processes and the application of CVD techniques to the synthesis of the new generation of thin films and nanomaterials. A description of particularly advanced techniques, such as spatial atomic layer deposition (SALD), direct liquid injection chemical vapor deposition (DLI-CVD), as well as electron cyclotron resonance chemical vapor deposition (ECR-CVD), is given in considerable detail. Moreover, the application of CVD techniques to materials of great interest, such as diamond-like carbon (DLC) coatings, graphene layers, and carbon nanofibers (CNFs), is described.

The book is divided into four sections, the first of which is dedicated to the description of the fundamentals of particularly advanced CVD techniques, while the following sections are devoted to the application of CVD techniques to specific types of materials: thin films, two-dimensional materials, and nanofibers. The first chapter describes the SALD technique, a variation of ALD in which precursors are continuously supplied in different locations and kept apart by an inert gas region or zone. Film growth is achieved by exposing the substrate to the locations containing the different precursors. Because the purge step is eliminated, the process becomes faster, more versatile, easier, and cheap to scale up. In addition, it can be performed at ambient pressure and even in the open air while not compromising deposition rate.

The second chapter is devoted to DLI-CVD, which, enabling the usage of solid and liquid precursors, has proven to be one of the most versatile CVD processes to meet the requirements for industrial application. The requirements to be met by the precursors suitable for DLI-CVD, the different classes of available precursors, as well as the models used to describe the evaporation process are overviewed in the chapter. Then, different liquid delivery devices used in DLI-CVD, such as capillary tubes, syringes, aerosol delivery systems, and valves, are reviewed in detail.

The third chapter deals with a general description of the CVD growth of DLC thin films, which are known and used in various branches of technology and medicine. A review of the most important growth methods is provided, describing the role of the physicochemical processes involved in the different techniques.

In the fourth chapter, research concerning CVD diamond coatings post-treated by plasma oxidation, with the aim of allowing the recycling of an original tungsten carbide with cobalt mother tool substrate, is presented. A developed radio-frequency/direct current plasma oxidation system is stated together with a hollow cathode device to intensify the oxygen ion and electron densities with quantitative plasma diagnosis equipment. Plasma oxidation ashing conditions are optimized by this quantitative diagnosis toward the perfect ashing of diamond films with fewer residuals and less tool edge damage. The geometric effect of tool teeth structure on this ashing process is discussed by *in situ* monitoring of plasma, and an engineering solution to this ashing process is proposed for industrial applications.

The fifth chapter presents a review of the preparation methods of high-quality graphene developed in recent years, considering the effect of experimental parameters, challenges of preparation of high-quality graphene, and timeline of graphene production. Finally, development trends and application prospects of CVD graphene are described.

The sixth chapter deals with the atmospheric pressure chemical vapor deposition (APCVD) of graphene, which is one of the best synthesis methods due to very low diffusivity coefficient and a critical step for graphene-based device fabrication. High-temperature APCVD processes are being recognized in many technologies, such as solid-state electronic devices, high-quality epitaxial semiconductor films, as well as metal oxide semiconductor transistors. In this chapter, recent advances in APCVD-based graphene synthesis and related applications are addressed.

The seventh chapter deals with helical carbon nanofibers (HCNFs), including carbon nanocoils (CNCs), carbon nanotwists, and multiwalled carbon nanotubes, which can be synthesized by CVD. HCNFs are predicted to have a high mechanical strength and hence are expected to have a use in nanodevices such as electromagnetic wave absorbers and sensors. For nanodevice applications, it is necessary to synthesize HCNFs in high yield and purity. The chapter focuses on CNCs, and describes its history, expected application, and synthesis method. Finally, it introduces the authors' recent studies on the improvement of the purity of CNCs by improving CVD conditions.

Prof. Pietro Mandracci
Department of Applied Science and Technology
Politecnico di Torino
Torino, Italy

Section 1

CVD Fundamentals

Chapter 1

Spatial Atomic Layer Deposition

David Muñoz-Rojas, Viet Huong Nguyen,
César Masse de la Huerta, Carmen Jiménez and
Daniel Bellet

Additional information is available at the end of the chapter

http://dx.doi.org/10.5772/intechopen.82439

Abstract

In conventional atomic layer deposition (ALD), precursors are exposed sequentially to a substrate through short pulses while kept physically separated by intermediate purge steps. Spatial ALD (SALD) is a variation of ALD in which precursors are continuously supplied in different locations and kept apart by an inert gas region or zone. Film growth is achieved by exposing the substrate to the locations containing the different precursors. Because the purge step is eliminated, the process becomes faster, being indeed compatible with fast-throughput techniques such as roll-to-roll (R2R), and much more versatile and easier and cheap to scale up. In addition, one of the main assets of SALD is that it can be performed at ambient pressure and even in the open air (i.e., without using any deposition chamber at all), while not compromising the deposition rate. In the present chapter, the fundamentals of SALD and its historical development are presented. Then, a succinct description of the different engineering approaches to SALD developed to date is provided. This is followed by the description of the particular fluid dynamics aspects and the engineering challenges associated with SALD. Finally, some of the applications in which the unique assets of SALD can be exploited are described.

Keywords: chemical vapor deposition, spatial atomic layer deposition (SALD), atmospheric pressure, in-line processing, thin films, transparent conductive materials, fluid dynamics

1. Introduction

The ALD technique was patented in 1977 by Suntola and Antson [1]. It is based on the self-terminating, surface-limited reactions of volatile precursor molecules with a substrate. As a result, ALD offers unique assets, namely, precise film thickness control to the nanometer, high-quality materials even at low temperature (thanks to the specific highly reactive nature of ALD precursors), and crack-free, compact and conformal film deposition even over high-aspect-ratio features. In this first patent, both the temporal and the spatial approaches were already proposed. In the former, precursors are injected in consecutive pulses separated by purge steps (and thus separated in time) [2–5]. In the later, precursors are supplied in different locations and it is the substrate that moves from one location to another (thus separated in space, see **Figure 1**) [6–9].

In all the enclosures proposed in Suntola's first patent, the reaction takes place inside deposition chambers that operate under vacuum. Curiously enough, although the three first reactors proposed by Suntola were spatial, followed by two temporal reactors (**Figure 2**), ALD has traditionally been developed only based on temporal approaches, both industrially and at the laboratory level, until recently. Some years afterward, in a new patent from 1983, Suntola introduced the idea of using an inert gas flow to separate the different precursors, as an alternative to purging under vacuum [10]. He again applied this principle to both temporal and spatial reactors (**Figure 3**). This was supported by an analytical study of precursor diffusion across the inert gas flow as a function of reactor design and gas flow rate and pulse time. The analysis showed that it was possible to conceive SALD reactors that could operate at ambient pressure without the need of unrealistic inert gas barrier flows, making atmospheric pressure SALD (AP-SALD) possible.

Thus, from the chemical point of view, spatial ALD is equivalent to temporal ALD, and therefore, self-terminating, surface-limited reactions occur. As a consequence, SALD also offers a highly precise growth per cycle, the ability to conformally coat high-aspect-ratio features, and the possibility to deposit high-quality films at lower temperatures than with chemical vapor deposition (CVD). But in addition to retaining the unique assets of ALD, SALD can be up to two orders of magnitude faster. Finally, the possibility to perform SALD at atmospheric pressure (AP-SALD), and even in the open air (without using any deposition chamber at all), makes it cheaper and easier to scale up since complex and expensive vacuum processing is not required.

Although the first patent involving atmospheric-pressure SALD dates from 1983, the first report applying the SALD approach was published only in 2004 [11]. After this initial report,

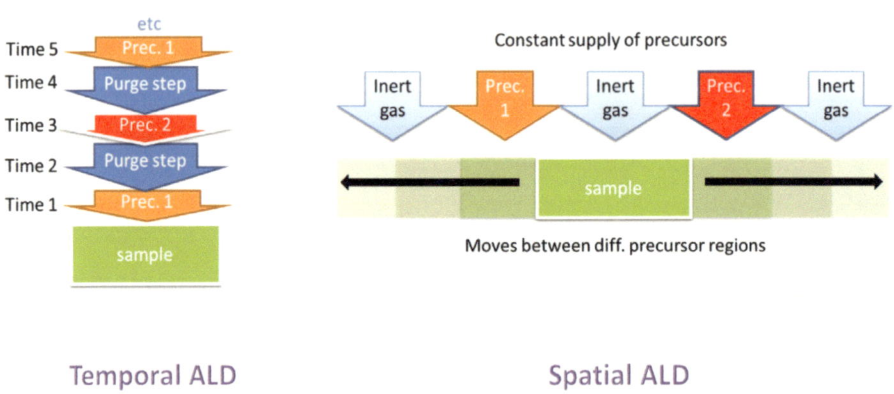

Figure 1. Schema of conventional ALD (left) and of SALD (right).

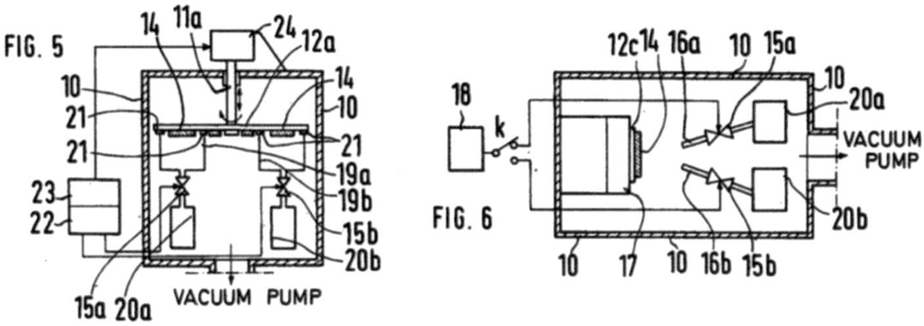

Figure 2. Two of the enclosures proposed by Suntola in his patent from 1977. Left: spatial approach and right: temporal approach.

the SALD field has experienced a significant growth, as shown by the increase in the number of publications dealing with SALD. In addition, many patents have been published since 1983 due to the appropriateness of the AP-SALD approach for application in industry and mass production. The first one by D. Levy (from Kodak) [12] describes a close-proximity, open-air reactor (see Section 2). Rather quickly, SALD has indeed reached industrial commercialization, and several companies are currently developing, fabricating, and/or selling SALD equipment, both for industrial production and for academic research [13].

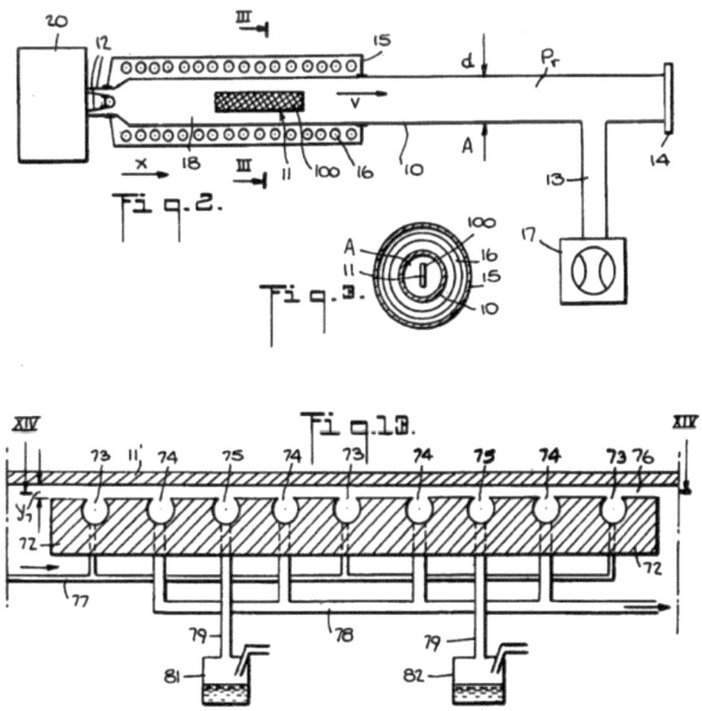

Figure 3. Two of the enclosures proposed by Suntola in his patent from 1983. Top: temporal approach and bottom: spatial approach.

2. Engineering approaches to SALD

The spatial approach has proven to be a very versatile one from the engineering point of view since many reactors have been reported to date. Nonobe et al. reported a spatial horizontal quartz hot-wall type reactor for the growth of HfO_2 films on 10 × 10 mm silicon substrates [11]. $HfCl_4$ (evaporated at 433 K) and O_2 are introduced in opposite sides of the reaction zone using a purified N_2 gas flow. A purified N_2 gas curtain is used to prevent precursors' mixing. Finally, the substrate is oscillated between the different zones by a computer-controlled system. Despite the spatial distribution of the precursors, the precursors are supplied in a temporal fashion, since pulses are delivered to the substrate once it is in place.

Since in most systems, the precursors are continuously fed into the reactor, SALD has also been referred to as continuous ALD. This is the case of the R2R reactor designed by Lotus Applied Technology [14]. In it, a web substrate is moved between reactor zones containing the different precursors, which are separated by a purge zone (**Figure 4**). Differential pressure and pumping are used to prevent precursor migration into the purge zone. Process pressure is generally about 2 mbar. Web speed can reach tens of meters per minute. The reactor has also been configured for plasma-based ALD mode since at high web speeds, it is found that water desorption is not fast enough, thus producing an anomalous growth (see Section 3 for more details).

The ASTRal group from the Lappeenranta University of technology developed a circular reactor for flexible substrates [15]. The design consists of a cylindrical drum to which the flexible substrate is attached. It then rotates inside a reaction chamber containing different precursors and purge zones. The process is equivalent to conventional ALD since the obtained growth per cycle saturates as the precursor flow rate is increased, reaching approximately 1 Å/cycle for the trimethylaluminum (TMA)/water process for the deposition of Al_2O_3 at 100 °C. Prof. Steven M. George's group from the University of Colorado has also developed a modular rotating cylindrical reactor [16]. The design is based on two concentric cylinders. The outer cylinder is fixed and contains several slits that can accept a wide range of modules that attach from the outside (**Figure 5**). The modules can easily move between the various slot positions to perform precursor dosing, purging, or pumping. The inner cylinder rotates with the flexible

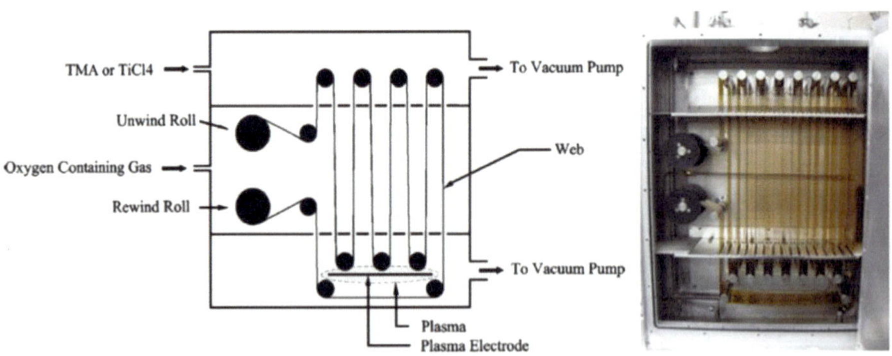

Figure 4. Low-pressure SALD reactor proposed by Lotus Applied Technology. Reproduced from Dickey, E. et al. J. Vac. Sci. Technol. A 30, 021502 (2012), with the permission of the American Vacuum Society.

http://dx.doi.org/10.5772/intechopen.82439

substrate passing underneath the various spatially separated slits in the outer cylinder. With this reactor, Al_2O_3 films can be grown at a rate of 2 Å/s, for rotation speeds of 175 RPM. Higher deposition rates can even be achieved by adding more modules.

Levy et al. from Kodak designed a very neat approach to AP-SALD in 2008 by [17]. In their approach, an injector manifold head is used to supply the different precursors along parallel channels. The precursors are kept separated by adjacent channels supplying an inert gas flows. Effective precursor separation is attained with practical gas flows (up to thousands of sccm (standard cubic centimeters per minute)) by placing the substrate close to the deposition head (around 50 microns). A relative motion between the head and the substrate replicates the ALD cycles yielding film growth (**Figure 6**). As a result, the system works at atmospheric pressure and even operates in the open air, that is, without using a deposition chamber. In the first model, the injector head was placed on top of the substrate. Later, a different design by Kodak was made in which the head lays at the bottom (the outlets facing upward), with the substrate oscillating on top. Later, other groups, such as the Laboratoire des Matériaux et du Génie Physique (LMGP) in Grenoble (**Figure 6**), have designed similar "close-proximity" SALD setups [18, 19]. In situ monitoring of film thickness has been performed by Yersak et al. taking advantage of the open-air environment offered by such AP-SALD approach. In their work, a reflectometer is placed in series with the injection module in order to monitor film thickness while depositing on a web.to (**Figure 7**) [20].

The close proximity approach, and the effective precursor separation that it provides, has been used in other reactors. TNO (Netherlands Organisation for Applied Scientific Research) has developed a cylindrical reactor consisting of separate zones exposing the precursors one by one to a circular substrate that moves underneath the precursor injector [21]. Between and around the reaction zones, shields of inert gas separate the precursor flows. Under the right operating conditions, these gas shields act as gas bearings, facilitating virtually frictionless movement between reactor and substrate. The group in TNO has also used this type of reactor to do plasma-activated SALD [22, 23]. The same group has also developed an R2R type of reactor based on the close proximity approach similar to the one developed by George's

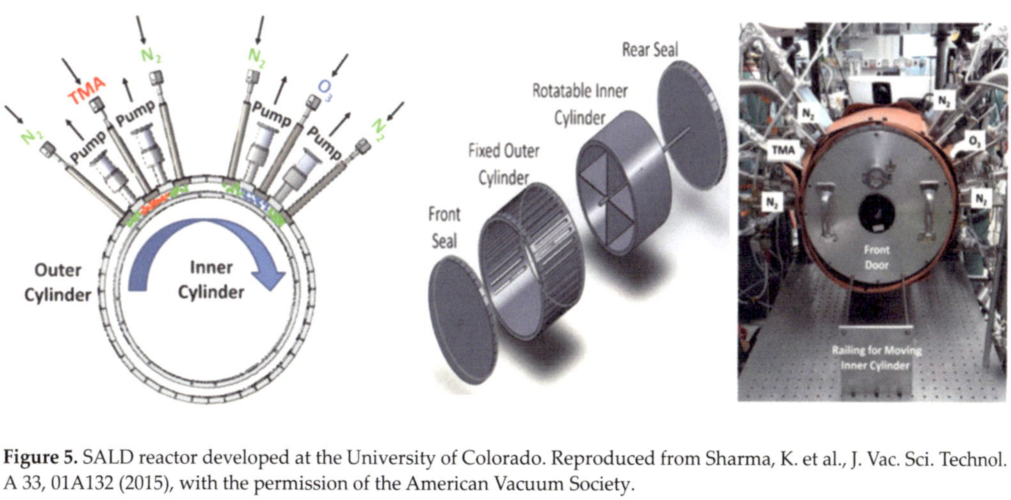

Figure 5. SALD reactor developed at the University of Colorado. Reproduced from Sharma, K. et al., J. Vac. Sci. Technol. A 33, 01A132 (2015), with the permission of the American Vacuum Society.

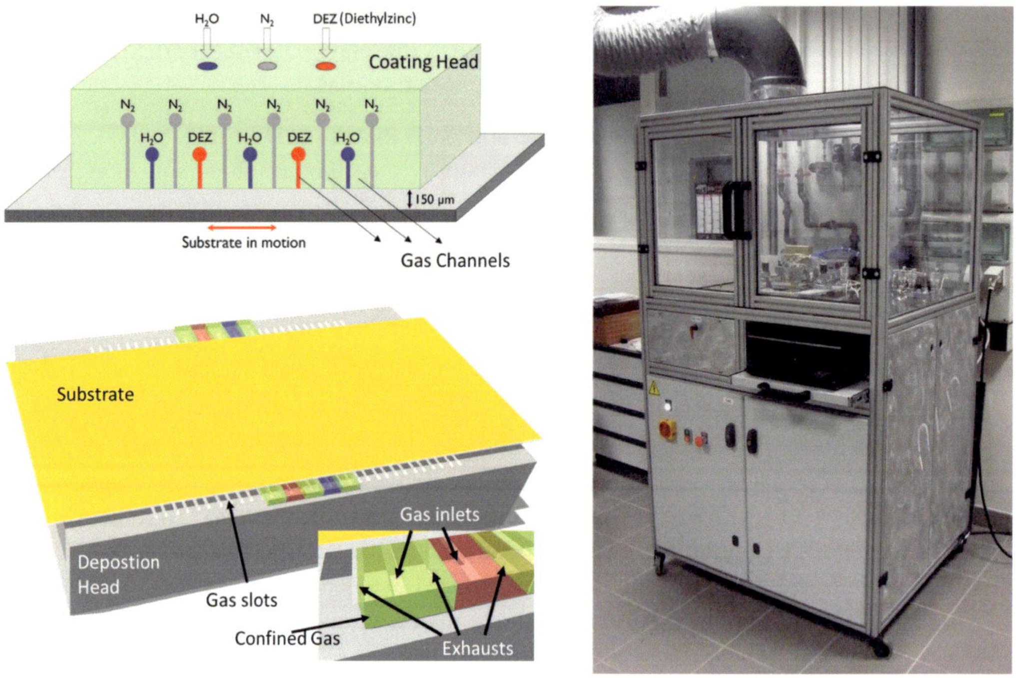

Figure 6. Close-proximity SALD approaches. Left: Schemes of the initial SALD systems reported by Kodak. Right: system developed at the LMGP.

group in Colorado (described above), but in this case, the gas-bearing principle is used (the web floated onto the gas flows coming out of the inner cylinder). There are indeed several patents concerning substrate floatation [24], and this approach has been used for the Solaytec and Levitech commercial systems (see below).

Finally, Ruud van Ommen et al. from the Delft University of Technology have developed a tubular SALD reactor capable of coating nanoparticles thanks to pneumatic transport (**Figure 8**) [25].

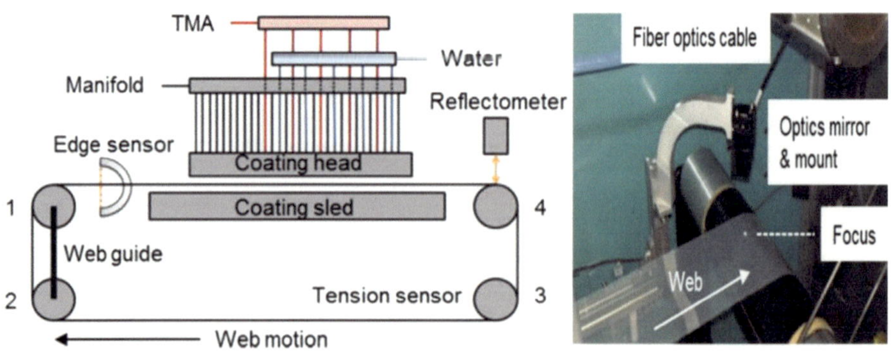

Figure 7. Implementation of in situ film thickness monitoring to an in-line AP-SALD reactor. Reproduced from Yersak, A. S. eta l., J. Vac. Sci. Technol. A 32, 01A130 (2014), with the permission of the American Vacuum Society.

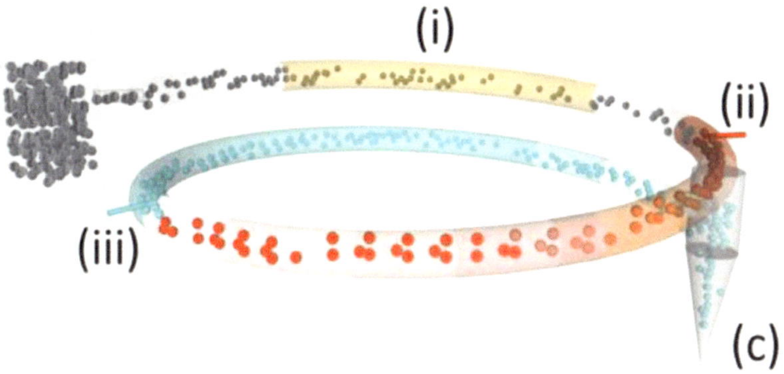

Figure 8. Tubular SALD reactor for coating nanoparticles. Reproduced from van Ommen, J. R. et al., J. Vac. Sci. Technol. A 33, 021513 (2015), with the permission of the American Vacuum Society.

They have developed a 27-m long, 4-mm internal diameter tube reactor, and the substrate nanoparticles (TiO_2 in this case) are fed at about 1 g min^{-1} into the tube from a vessel in which they are suspended in an upward N_2 flow (typically called a fluidized bed). The particles are carried inside the tube at velocities of 5 ms^{-1}. Different sections of the reactor tube contain the different precursor and inert gas purge zones.

The high throughput and scalability of AP-SALD has already been harnessed, and industrial and laboratory equipment is already commercialized. Solaytec (www.solaytec.com), a spin-off from TNO, developed a linear modular close-proximity reactor specially designed to deposit Al_2O_3 passivation layers on silicon solar cells. Levitech (www.levitech.nl) proposes a similar approach. Beneq has as a well-developed, large area R2R industrial coating SALD system (WCS 600), along with the laboratory-scale SALD systems (TFS 200R, R11). A large-scale, large area SALD reactor has recently been presented for the coating of large flat substrates. Applied Materials and Jusung Engineering also propose SALD reactors. SALD is also being used to functionalize commercial products (e.g., plastics) using home-built equipment.

3. SALD fluid dynamics and modeling

From the chemical point of view, SALD is equivalent to ALD and in order to have layer-by-layer surface limited growth, the surface to be coated must be saturated, and any excess precursor that stays physisorbed or laying above the surface should be removed in the purge zone. Since in SALD, the precursors are supplied continuously, efficient separation is a key issue and the reactors must be designed so that the inert flow barriers are effective. As we have seen in the previous section, this has been achieved using various engineering approaches. In order to evaluate the effectiveness of a particular reactor and the effect of the different conditions on the process, modeling is frequently used, as shown below.

The first main advantage of SALD with respect to ALD is that processing can be much faster. The maximum deposition rate that is achievable depends of course on the type of reactor used, but ultimately, it always depends on the properties of the different precursors and the reaction rate (i.e., reaction kinetics). The time that the substrate is in contact with the precursor flow must be long enough for achieving a complete reaction (i.e., surface saturation). If we take the reaction between TMA and water as an example, the time scale for the TMA half-reaction can be estimated by

$$t_{HR} = \frac{A\chi\, d_{mono}\, \rho_{Al_2O_3}}{c_{gs}\left(M_{Al_2O_3}/M_{TMA}\right) \varnothing_{TMA}} \tag{1}$$

with A being the area available for deposition, χd_{mono} being the self-limiting film thickness, ρ_{Al2O3} the density of the deposited layer, c_{gs} the stoichiometric coefficient, M_{Al2O3} and M_{TMA} are the molar masses of Al_2O_3 and TMA, and $\varnothing_{TMA}$ the precursor mass flow rate. Poodt et al. used this approach to estimate a time scale of a few milliseconds for their reactor [21].

Substrate speed thus needs to be adjusted in order to ensure surface saturation. At the same time, speed may also be limited by the desorption kinetics of precursor molecules physisorbed on the substrate. Water, for example, is known to desorb increasingly slower as temperature diminishes (meaning than in conventional ALD, purging steps become much longer, i.e., 30 s and even minutes), and plasma oxidation is a better choice for low temperatures [14]. The rate of desorption is proportional to the surface concentration, Cs, and on the probability that water molecules desorb. Taking the binding energy of absorbed molecules from the Boltzmann equation, the desorption rate per molecule can be expressed as:

$$R_d = L exp\left(-\frac{qE}{kT}\right) \tag{2}$$

where L is a constant including the molecular vibrational frequency, E is the molecule binding energy (eV), q is the electronic charge, k is the Boltzmann's constant (JK^{-1}), and T is the temperature (K). The rate of molecules accumulation is then given as:

$$dCs/dt = Is - CsRd \tag{3}$$

where I is the impingement rate of water molecules and s is the sticking coefficient. I can be expressed as

$$I = \frac{P N_A}{\sqrt{2\pi MkT}} \tag{4}$$

where N_A is the Avogadro's number, M is the molar mass (kg), and P is the partial pressure of the water vapor. Eq. (3) can thus be solved as

$$C_s = Is\tau\left(1 - \exp\left(-\frac{t}{\tau}\right)\right) \tag{5}$$

where

$$\tau = \frac{1}{R_d} = \frac{1}{L}\exp\left(\frac{qE}{kT}\right) \tag{6}$$

is a time constant. So, the excess of water molecules absorbed on the substrate as a function of time spent in the precursor zone can be approximated as

$$Q_0 = Is\tau(1 - exp(-1/\tau))\ for\ t > t_m \tag{7}$$

where t_m is the time required for having a monolayer (thus $Q_0 = 0$ for $t < t_m$). From the above discussion, the amount of excess growth (i.e., anomalous CVD growth due to remaining physisorbed water in the substrate) is obtained by combining the water accumulation on the surface while the substrate is in the precursor zone and the desorption taking place in the purging zone:

$$\Delta G = bIs\tau(1 - \exp(-t/\tau))\exp(-t/\tau) \tag{8}$$

where b is a constant. Thus, temperature, precursor concentration, and speed need to be carefully adjusted in order to ensure minimal excess of absorbed molecules and effective purging. Maydannik et al. used this analytical approach to fit anomalous Al_2O_3 growth obtained experimentally between 100 and 150 °C [26].

The second advantage of SALD is that it can be easily performed at atmospheric pressure and in the open air. For that, the precursors need to be isolated from the atmosphere by the inert gas flow. In its patent from 1983 [10], Suntola developed the equations allowing to estimate the flows that would be required in order to ensure precursor separation and isolation from the ambient. The equations can be applied to temporal ALD reactors and spatial ones, after proper modification taking into account the particular design of each reactor. The equations developed allowed him to predict the viability of atmospheric pressure SALD without the need for unrealistically high inert gas flows.

An important parameter affecting the SALD process is reaction temperature, since it can affect effective precursor separation and the kinetics of the different reactions and processes (such as desorption) taking place. As a consequence, the effective precursor separation and the maximum achievable speed for a particular reactor and reaction can vary for different deposition temperatures. For example, Pan et al. have performed the simulations of a reactor based on the manifold head injector approach [27]. They have used the following set of equations for their simulations:

$$\frac{d\rho}{dt} + \nabla \cdot (\rho\vec{V}) = 0 \tag{9}$$

$$\frac{d}{dt}d(\rho\vec{V}) + \nabla \cdot (\rho\vec{V}\vec{V}) = -\nabla P + \nabla \cdot \tilde{\tau} + \rho\vec{g} + \vec{F} \tag{10}$$

$$\frac{d}{dt}d(\rho\, c_i) + \nabla \cdot (\rho\, c_i\vec{V}) = -\nabla \cdot \vec{J}_{m,i} + R_i \tag{11}$$

$$\frac{d}{dt}d(\rho E) + \nabla\left[\vec{V}(\rho E + P)\right] = \nabla \cdot \left[k\,\nabla\, T - \sum_i h_i \vec{J}_{h,i} + (\tilde{\tau} \cdot \vec{V})\right] \tag{12}$$

where ρ is the density, $\vec{V}$ is the velocity vector, P is the static pressure, $\rho\vec{g}$ and $\vec{F}$ are the gravitational body force and external body forces, respectively, $\tilde{\tau}$ is the stress tensor, ci is the local molar fraction of species i, R_i is the net rate of production of species i by chemical reaction, $\vec{J}_{m,i}$ is the mass diffusive flux of mixture species i, k is the material thermal conductivity, hi is the enthalpy of mixture species i, and $\vec{J}_{h,i}$ is the energy diffusive flux of mixture species i.

The results of the modeling show that temperatures higher than 250 °C accelerate the diffusive mass transport, thus inducing precursor intermixing. On the other hand, high temperature also increases the film deposition rate and results in a higher saturated growth per cycle (GPC) level. The simulation also showed that the chemical deposition process is highly affected by the flow and concentration of precursors. Surface reaction kinetics of the in-line spatial ALD are thus a function of deposition temperature, flow conditions, reactive surface sites, and precursor distributions.

In another example, Deng et al. have presented a numerical model for atmospheric SALD process, based again on an in-line (i.e., manifold injector head) reactor [28]. The effect of inert gas flow rate, carrier gas flow rate, and precursor concentration (i.e., partial pressure or mass fraction) is addressed. The inert gas flow rate obtained from the model agrees well with the experimental values. Simulation results show that the precursor concentration is the determinant factor governing minimal residual time; the optimum precursor usage is inversely proportional to the carrier gas flow rate. Thus, for a constant carrier gas flow, the optimal precursor usage and the precursor mass fraction form a monotonic decreasing relationship. The gap between the gas head and the substrate, regardless of its specific structures in any SALD system, leads to a low Péclet number, thus implying that precursor diffusion plays a more important role than convection.

Finally, due to the relative movement between the substrate and the different gas zones, the possibility of carrying some precursor molecules from one precursor zone to the second precursor zone across the inert gas barrier is not negligible. This precursor entrainment is very much related to the sample speed and to the inert gas flow rate. The movement between a surface and a gas gives rise to a boundary layer at the surface having a thickness inversely proportional to the square root of the relative velocity [15, 26]. The relationship between layer thickness, δ, and substrate speed (Vs) can be expressed as:

$$\delta \sim \sqrt{\frac{1}{V_s}} \tag{13}$$

The dependence of the concentration of precursor in the boundary layer is proportional to δ, which in turn is proportional to the square root of the substrate residence time (residence time being inversely proportional to substrate velocity). Since precursor concentration falls from maximum to 0 once the substrate enters a purge zone, the total quantity of precursor in the boundary layer can be expressed as $G = (1/2)C_s\delta$, where C_s is the concentration at the substrate surface. G diminishes as the substrate progresses in the purge zone. The flux of precursor out of the boundary layer will depend on the concentration gradient:

$$F(t) = -D\frac{dC}{dx} \approx -D\frac{C_s}{V_s} \approx -D\frac{2G}{V_s^2} \tag{14}$$

where D is the diffusion coefficient of the precursor. A similar approach to the one used above to evaluate the excess growth due to excess precursor physisorption can be used here to evaluate anomalous growth rate due to precursor entrainment.

For close-proximity systems, the gap between injector and sample is very important in order to ensure no mixing of the precursors in the gas phase. Nevertheless, for systems for which the gap can be adjusted mechanically, there is also the possibility to perform the deposition in a pseudo-CVD (or spatial CVD, SCVD) mode, that is, allowing a partial mixing of precursors immediately above the substrate. For samples not having complicated or high-aspect-ratio features, not having CVD is not an issue and the obtained films are still dense, uniform, and the thickness obtained is proportional to the number of cycles [29]. In order to control the deposition mode, computational fluid dynamic approaches (CFD) can be used to simulate the process (**Figure 9**). It is also important to consider that the properties and the morphologies of the films obtained using the two different modes (SALD vs. SCVD) are usually different [9].

The possibility of having precursor molecules being entrained from the region where they are being injected to the region where the other precursor is being injected has also to be considered. This is due to the relative movement of the substrate and the injector or precursor zones, and the existence of a boundary layer [15, 19, 20, 30–32]. Again, CFD approaches can be used to predict whether precursor entrainment is to be expected for a set of conditions with a specific SALD system. For example, while many close-proximity systems operate with gap

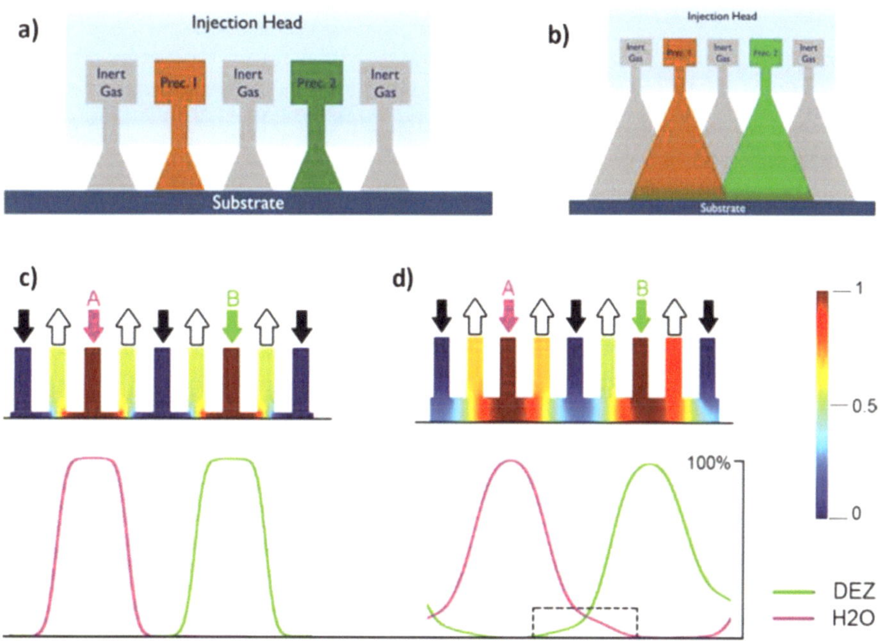

Figure 9. Comparison between SCVD and SALD modes. Scheme of a close-proximity head having two precursor channels and different gaps between head and sample: (a) 150 μm gap, no mixing in the gas phase, (b) 700 μm gap, resulting in precursor mixing above the sample surface, and (c) and (d) show a CFD stationary simulation of the two different modes. A whole representation of the head was used to simulate the behavior of the gases, but only two precursor channels are shown. A cross section of the head with precursor, inert and evacuation channels is shown in which the normalized concentration of the different precursors is given. Below, normalized precursor concentration profiles calculated along the substrate line are shown.

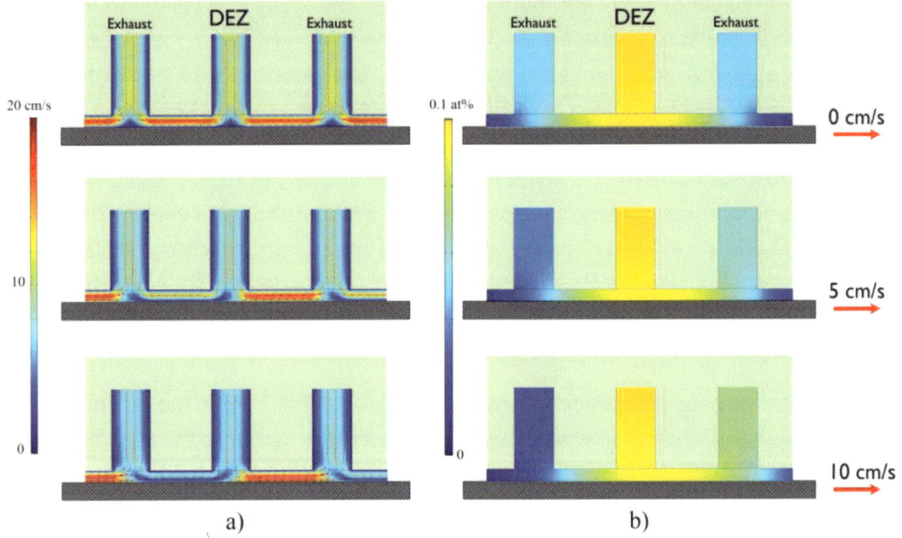

Figure 10. Spatial distribution of (a) precursor velocity and (b) precursor concentration above the substrate for the deposition of ZnO from diethyl zinc (DEZ). Here, only the cross-section views including a DEZ inlet and two exhaust inlets are shown. The blue color corresponds to low velocity/precursor concentration, while red or yellow color corresponds to high velocity/precursor concentration.

sizes in the order of 100 μm, Pan et al. have shown that injector-substrate gaps of as a much as 1.5 mm can be used without having precursor cross-talk [27]. **Figure 10** shows a simulation of the effect of substrate speed in the gas velocity and precursor concentration for the close-proximity SALD system at LMGP [9, 33]. As it can be seen, with the conditions used for the simulation, no precursor entrainment takes place between different precursor regions. The evaluation of precursor entrainment has also been monitored experimentally [30]. Finally, it has recently been demonstrated that atmospheric pressure SALD can be used to coat porous substrates with high throughput as long as high precursor partial pressures and molar flows can be reached [34].

4. Taking advantage of SALD

To date, several materials including intrinsic, doped and mixed oxides, metals and recently organic coatings, have been produced using SALD reactors, as detailed in **Table 1**. The applications range from active and passive components for thin film transistors (TFTs), barriers passivation layers to active components for new-generation solar cells. Some examples are briefly described next.

The first application of SALD, from Kodak, involved the deposition of ZnO and Al_2O_3 layers as components in TFTs [17]. The same group has published different studies on this line, including the use of growth inhibitors for the deposition of patterned films, and thus showing that selected area deposition is also possible with spatial ALD [36, 39, 82, 83]. The group from TNO used their rotary reactor to deposit Al_2O_3 passivation layers for silicon solar cells, and the good

http://dx.doi.org/10.5772/intechopen.82439

Material		References
Intrinsic metal oxides	HfO_2	[11]
	Al_2O_3	[14–16, 19–21, 26, 31, 35–50]
	ZnO	[18, 36, 39, 45, 46, 51–67]
	SnO_x	[68, 69]
	TiO_2	[14, 18, 70–72]
	Cu_2O	[46, 73, 74]
	Nb_2O_5	[71]
	NiO_x	[91]
	ZrO_2	[92, 93]
	MoO_x	[94]
Doped oxides	ZnO:N	[46, 55, 63, 67]
	Al:ZnO	[36, 56, 75]
	$Zn_{1-x}Mg_xO$	[46, 64, 76]
	In:ZnO	[51, 77]
	ZnO:S	[58]
	$TiO_{2-x}Cl_{2x}$	[78]
Mixed oxides	IGZO	[79]
	CIGS	[80]
Metals	Ag	[23]
	Pt	[25]
Organic materials	Polyamide	[81]

Table 1. Materials deposited by SALD to date.

results prompted the creation of the spinoff company Solaytec (see above) [21]. The same group has dedicated efforts to deposit transparent conductive oxides (TCO) layers and more complex oxides for application in photovoltaic devices [51, 56, 84, 85]. Prof. Driscoll's group was the first one to apply the SALD technique for the deposition of active layers for new-generation solar cells [52, 71–74]. For example, Muñoz-Rojas et al. showed that a 15-nm thick amorphous TiO_2 layer can act as an efficient hole blocking layer in bulk heterojunction solar cells [72]. Thanks to the high quality offered by the SALD technique, an extremely thin (yet crack and pinhole free) layer can be used. And as a result, low deposition temperatures can be used, since the lower conductivity of amorphous TiO_2 is compensated by having a very thin layer. In addition, deposition rates are two orders of magnitude faster than for other low-temperature, atmospheric deposition methods (**Figure 11**). Such high throughput characteristic of SALD is of high relevance for organic-based PV technology. Recently, it has also been shown that SnO_x and ZnO films deposited by SALD can have a beneficial impact on the stability of hybrid perovskite solar cells [62, 69].

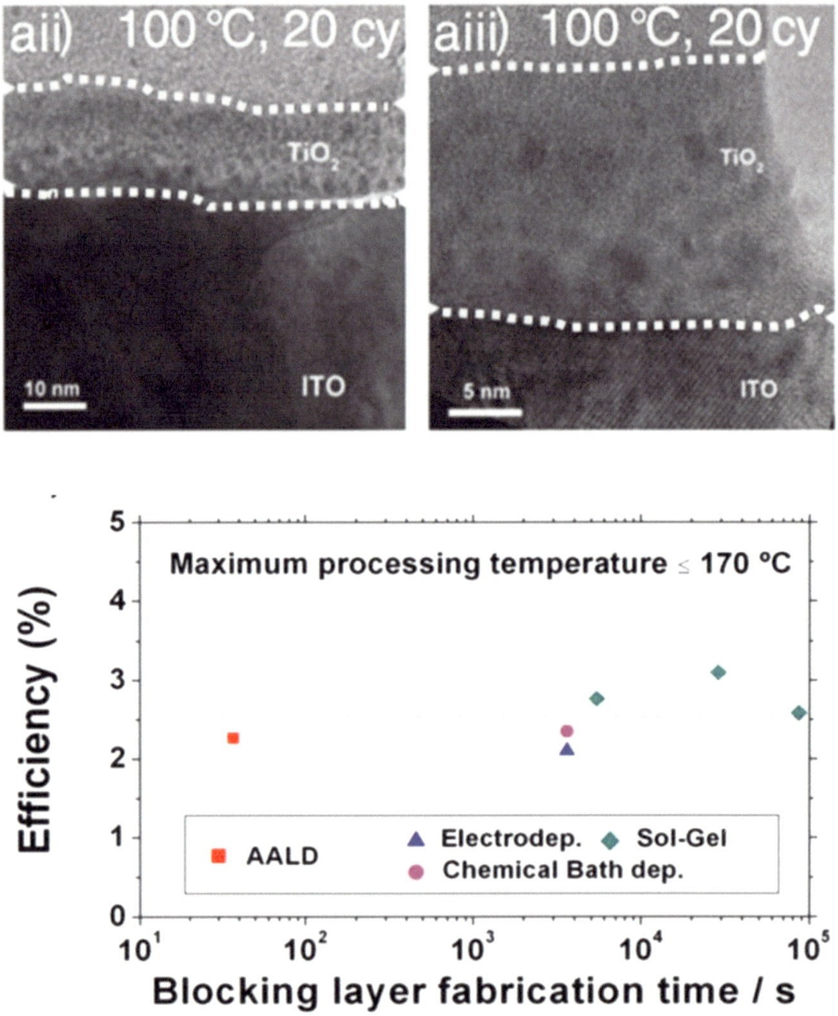

Figure 11. Top: TEM images of TiO_2 amorphous films deposited at 100 °C. Bottom: cell efficiency compared with equivalent cells from the literature in terms of blocking layer fabrication time [72].

SALD has also been used for LEDs, in some cases to deposit active ZnO layers for polymer and hybrid perovskite-based diodes [54, 76], and in another case using Al_2O_3 as permeation barrier for flexible organic LEDs [49]. Other authors have also demonstrated the suitability of SALD for depositing barrier and encapsulation layers both on plastic and paper substrates [14, 31, 40, 48].

K. Sharma et al. used their cylindrical reactor (the University of Colorado, see Section 2) to study the deposition of ZnO on 25 μm thick flexible anodized aluminum oxide (AAO) templates [86]. They used DEZ and ozone as precursors, and the reaction temperature was set at 50 °C. The results show that pores with 100 nm of diameter can be conformally coated for substrate speeds ≤10 RPM. This corresponds to an aspect ratio of 250, and the corresponding residence time is 48 ms. For faster substrate speeds or smaller pores, the films were not uniform along the pores. Coating of the AAO was used to evaluate the conditions for coating porous substrates. They then used optimized conditions to coat porous Li battery electrodes. The Al_2O_3 SALD coating used was observed to enhance the capacity stability of the electrodes. Other recent studies show the benefits of SALD for the field of energy storage [45, 78].

Chen et al. have developed a system in which two injector heads are combined for the deposition of ZnO/TiO_2 nanolaminates [18]. They show that nanolaminates with the same thickness and overall composition could be turned from crystalline to amorphous by increasing the number of bilayers (i.e., by reducing the thickness of each individual bilayer). Optical transmittance of the laminates increases in the visible range with the number of bilayers. The refractive index of laminates increases with decreasing bilayer thickness, thus demonstrating the possibility for tuning. The conductivity of nanolaminates is higher than for ZnO films due to an increase in carrier density. On the other hand, mobility decreases with respect to intrinsic ZnO layers.

Recent reports from Muñoz-Rojas and coworkers have shown that SALD is an ideal technique to enhance the stability of transparent electrodes based on metallic nanowires (MNWs). In these electrodes, metallic nanowires are deposited by different low-cost, scalable methods, such as spray coating, to obtain a random network. Above a certain areal mass density of the electrode (i.e., amount of nanowires), percolation of the nanowires occurs and the electrode becomes conductive. Yet, since most of the space is empty, the electrode remains transparent. The best electrodes based on AgNWs are comparable and even better than the state-of-the-art indium tin oxide (ITO, the most commonly used transparent conductive oxide) electrodes [87–89]. In addition, electrodes based on MNWs are flexible, which in combination with the soft deposition methods used to fabricate the networks, and make this technology

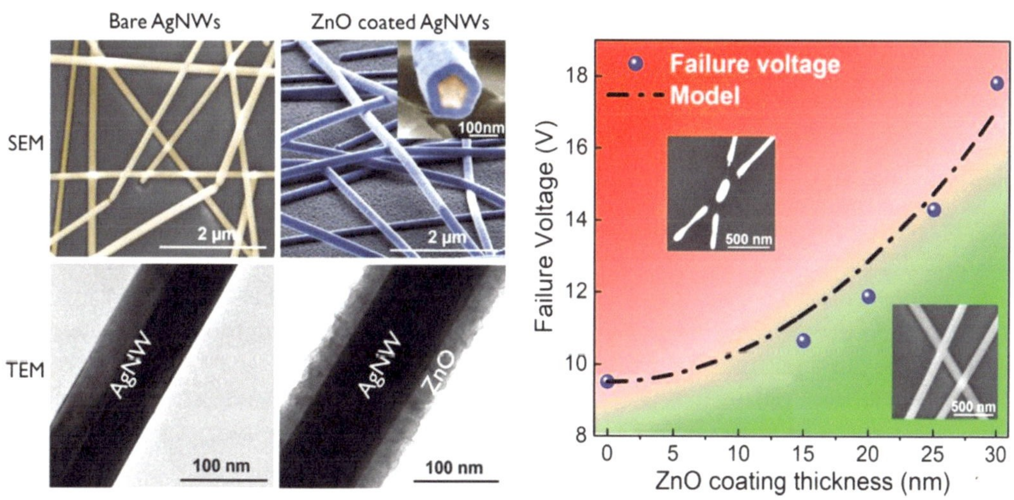

Figure 12. Left: SEM and TEM images of bare and ZnO-coated AgNWs. Right: failure voltage of AgNWs networks versus ZnO coating thickness. The insets correspond to SEM images of a network in which the wires have started spheroidizing due to thermal and electrical instabilities (top), and to a coated network in which the nanowires are intact after being subjected to electrical stress (bottom). The discontinuous line is obtained using a simple model that shows that the enhanced stability of the coated networks is due to hindered Ag atomic diffusion (see reference for details) [61].

compatible with plastic and other soft substrates. Finally, electrodes based on MNWs are more transparent to the infrared part of the spectrum than the transparent conductive oxides (TCO). Despite these advantages, electrodes made of MNWs suffer from thermal, electrical, and chemical instabilities. A solution to increase the stability of the nanowires is to coat them with a thin oxide film, which prevents the diffusion of metal atoms in the wires and the reaction of the same with the atmosphere. ALD is a perfect technique for that since the coatings obtained are dense and crack free. But conventional ALD is usually performed in vacuum, and growth rate is too low. SALD on the other hand can be processed in the open air and offers a high deposition throughput. The close-proximity system at the LMGP has thus been used to coat MNW electrodes. In a first example, ZnO coatings of different thicknesses were deposited on an AgNWs electrode. The oxide coating resulted in a better electrical and thermal stability of the electrodes, and the improvement observed was directly proportional to the thickness of the oxide layer (**Figure 12**) [61]. In another example, Al_2O_3 coatings were used to protect CuNWs electrode. The encapsulation of the wires with the Al_2O_3 layer prevented the oxidation of the wires upon heating in the atmosphere, which allowed the fabrication of stable transparent heaters [44].

5. Conclusions

Although already patented in 1977 at the same time than temporal ALD, spatial ALD has only been developed with the beginning of the twenty-first century. The possibility to perform ALD in much faster deposition rate, at atmospheric pressure and even in the open air has converted SALD in to a technique that is gaining much attention and momentum. SALD has also made the transition from laboratory to industrial scale, and different commercial systems are already available both for laboratory and production scale, while home-made systems are also used. SALD is a very flexible deposition technology that allows a high degree of design freedom, as shown by the increasing number of reactors being developed. Because in SALD, precursors are continuously being injected, efficient separation by the inert gas flow/zone needs to be ensured. The analytical study of fluid dynamics and modeling are thus commonly used during the design of reactors and to evaluate optimum deposition conditions.

Many materials have been already deposited using SALD. Initial works focused on Al_2O_3 and ZnO. Later on, other binary oxides such as Cu_2O, TiO_2, or Nb_2O_5 have been developed, together with the deposition of more complex oxides (including doping and mixes oxides), metals, and even organic films, in which is the first example of spatial molecular layer deposition (SMLD) [90]. The main applications of SALD to date have been for the deposition of components for TFTs, solar cells, and LEDs and for the deposition of barrier and encapsulation layers. SALD has been demonstrated on flexible substrates such as paper or plastic and even on features with high aspect ratios.

The combination of the unique assets of ALD with faster deposition rates and air processing, design flexibility, and easy scalability are expected to make SALD one of the main thin-film deposition techniques in the coming years.

Acknowledgements

David Muñoz-Rojas acknowledges funding through the Marie Curie Actions (FP7/2007-2013, grant agreement no. 631111). DMR also acknowledges funding from the European Union's Horizon 2020 FETOPEN-1-2016-2017 research and innovation programme under grant agreement 801464. The authors thank the Agence Nationale de Recherche (ANR, France) for funding via the projects INDEED (no. ANR-15-CE05-0019) and DESPATCH (no. ANR-16-CE05-0021). This project has received funding from the European Union's Horizon 2020 research and innovation program under grant agreement no. 801464. Viet Huong Nguyen thanks the "ARC Energy Auvergne-Rhône Alpes" for the economic support for Ph.D. grants. César Masse de la Huerta acknowledges and thanks the funding received by CONACYT, Mexico. This work is supported by the French National Research Agency in the framework of the "Investissements d'avenir" program (ANR-15-IDEX-02) through the project Eco-SESA. This project was also financially supported by the "Carnot Énergies du futur" Institute (Aldash Project).

Conflict of interest

The authors declare no conflict of interest.

Author details

David Muñoz-Rojas*, Viet Huong Nguyen, César Masse de la Huerta, Carmen Jiménez and Daniel Bellet

*Address all correspondence to: david.munoz-rojas@grenoble-inp.fr

Université Grenoble Alpes, CNRS, Grenoble INP, LMGP, Grenoble, France

References

[1] Suntola TS, Antson J, Method for Producing Compound Thin Films. US 4,058,430. 1977

[2] George SM. Atomic layer deposition: An overview. Chemical Reviews. 2010;**110**:111-131. DOI: 10.1021/cr900056b

[3] Leskelä M, Ritala M, Nilsen O. Novel materials by atomic layer deposition and molecular layer deposition. MRS Bulletin. 2011;**36**:877-884. DOI: 10.1557/mrs.2011.240

[4] Puurunen RL. Surface chemistry of atomic layer deposition: A case study for the trimethylaluminum/water process. Journal of Applied Physics. 2005;**97**:121301. DOI: 10.1063/1.1940727

[5] Knez M, Nielsch K, Niinistö L. Synthesis and surface engineering of complex nanostructures by atomic layer deposition. Advanced Materials. 2007;**19**:3425-3438. DOI: 10.1002/adma.200700079

[6] Poodt P, Cameron DC, Dickey E, George SM, Kuznetsov V, Parsons GN, et al. Spatial atomic layer deposition: A route towards further industrialization of atomic layer deposition. Journal of Vacuum Science & Technology A—Vacuum Surfaces and Films. 2012;**30**:010802. DOI: 10.1116/1.3670745

[7] Muñoz-Rojas D, MacManus-Driscoll J. Spatial atmospheric atomic layer deposition: A new laboratory and industrial tool for low-cost photovoltaics. Materials Horizons. 2014;**1**: 314-320. DOI: 10.1039/c3mh00136a

[8] Musselman KP, Uzoma CF, Miller MS. Nanomanufacturing: High-throughput, cost-effective deposition of atomic scale thin films via atmospheric pressure spatial atomic layer deposition. Chemistry of Materials. 2016;**28**:8443-8452. DOI: 10.1021/acs.chemmater.6b03077

[9] Muñoz-Rojas D, Nguyen VH, Masse de la Huerta C, Aghazadehchors S, Jiménez C, Bellet D. Spatial Atomic Layer Deposition (SALD), an emerging tool for energy materials. Application to new-generation photovoltaic devices and transparent conductive materials. Comptes Rendus Physique. 2017;**1**:1-10. DOI: 10.1016/j.crhy.2017.09.004

[10] Suntola TS, Pakkala AJ, Lindfors SG. Apparatus for Performing Growth of Compound Thin Films. US 4,389,973. 1983

[11] Nonobe S, Takahashi N, Nakamura T. Preparation of HfO_2 nano-films by atomic layer deposition using $HfCl_4$ and O_2 under atmospheric pressure. Solid State Sciences. 2004;**6**: 1217-1219. DOI: 10.1016/j.solidstatesciences.2004.07.030

[12] Levy DH. Process for Atomic Layer Deposition. US 7,413,982 B2. 2008

[13] Muñoz-Rojas D, Maindron T, Esteve A, Piallat F, Kools J, Decamps M. Speeding up the unique assets of atomic layer deposition. Materials Today Chemistry. 2018

[14] Dickey E, Barrow WA. High rate roll to roll atomic layer deposition, and its application to moisture barriers on polymer films. Journal of Vacuum Science & Technology A—Vacuum Surfaces and Films. 2012;**30**:021502. DOI: 10.1116/1.3678486

[15] Maydannik PS, Kääriäinen TO, Cameron DC. An atomic layer deposition process for moving flexible substrates. Chemical Engineering Journal. 2011;**171**:345-349. DOI: 10.1016/j.cej.2011.03.097

[16] Sharma K, Hall RA, George SM. Spatial atomic layer deposition on flexible substrates using a modular rotating cylinder reactor. Journal of Vacuum Science & Technology A—Vacuum Surfaces and Films. 2015;**33**:01A132. DOI: 10.1116/1.4902086

[17] Levy DH, Freeman D, Nelson SF, Cowdery-Corvan PJ, Irving LM. Stable ZnO thin film transistors by fast open air atomic layer deposition. Applied Physics Letters. 2008;**92**:192101. DOI: 10.1063/1.2924768

[18] Chen R, Lin J-L, He W-J, Duan C-L, Peng Q, Wang X-L, et al. Spatial atomic layer deposition of ZnO/TiO_2 nanolaminates. Journal of Vacuum Science & Technology A—Vacuum Surfaces and Films. 2016;**34**:051502. DOI: 10.1116/1.4955289

[19] Ryan Fitzpatrick P, Gibbs ZM, George SM. Evaluating operating conditions for continuous atmospheric atomic layer deposition using a multiple slit gas source head. Journal of Vacuum Science & Technology A—Vacuum Surfaces and Films. 2012;**30**:01A136. DOI: 10.1116/1.3664765

[20] Yersak AS, Lee YC, Spencer JA, Groner MD. Atmospheric pressure spatial atomic layer deposition web coating with in situ monitoring of film thickness. Journal of Vacuum Science & Technology A—Vacuum Surfaces and Films. 2014;**32**:01A130. DOI: 10.1116/1.4850176

[21] Poodt P, Lankhorst A, Roozeboom F, Spee K, Maas D, Vermeer A. High-speed spatial atomic-layer deposition of aluminum oxide layers for solar cell passivation. Advanced Materials. 2010;**22**:3564-3567. DOI: 10.1002/adma.201000766

[22] Poodt P, Kniknie B, Branca A, Winands H, Roozeboom F. Patterned deposition by plasma enhanced spatial atomic layer deposition. Physica Status Solidi (RRL)—Rapid Research Letters. 2011;**5**:165-167. DOI: 10.1002/pssr.201004542

[23] van den Bruele FJ, Smets M, Illiberi A, Creyghton Y, Buskens P, Roozeboom F, et al. Atmospheric pressure plasma enhanced spatial ALD of silver. Journal of Vacuum Science & Technology A—Vacuum Surfaces and Films. 2015;**33**:01A131. DOI: 10.1116/1.4902561

[24] Levy DH. Deposition System and Method Using a Delivery Head Separated Froma Substrate by Gas Pressure. US 2009/0130858, US 2010/0311130 Al. 2009

[25] van Ommen JR, Kooijman D, de Niet M, Talebi M, Goulas A. Continuous production of nanostructured particles using spatial atomic layer deposition. Journal of Vacuum Science & Technology A—Vacuum Surfaces and Films. 2015;**33**:021513. DOI: 10.1116/1.4905725

[26] Maydannik PS, Kaariainen TO, Cameron DC. Continuous atomic layer deposition: Explanation for anomalous growth rate effects. Journal of Vacuum Science & Technology A—Vacuum Surfaces and Films. 2012;**30**:01A122. DOI: 10.1116/1.3662861

[27] Pan D, Jen T-C, Yuan C. Effects of gap size, temperature and pumping pressure on the fluid dynamics and chemical kinetics of in-line spatial atomic layer deposition of Al_2O_3. International Journal of Heat and Mass Transfer. 2016;**96**:189-198. DOI: 10.1016/j.ijheatmasstransfer.2016.01.034

[28] Deng Z, He W, Duan C, Chen R, Shan B. Mechanistic modeling study on process optimization and precursor utilization with atmospheric spatial atomic layer deposition. Journal of Vacuum Science & Technology A—Vacuum Surfaces and Films. 2016;**34**:01A108. DOI: 10.1116/1.4932564

[29] Hoye RLZ, Muñoz-Rojas D, Musselman KP, Vaynzof Y, MacManus-Driscoll JL. Synthesis and modeling of uniform complex metal oxides by close-proximity atmospheric pressure chemical vapor deposition. ACS Applied Materials & Interfaces. 2015;**7**: 10684-10694. DOI: 10.1021/am5073589

[30] Maydannik PS, Kääriäinen TO, Lahtinen K, Cameron DC, Söderlund M, Soininen P, et al. Roll-to-roll atomic layer deposition process for flexible electronics encapsulation applications. Journal of Vacuum Science and Technology A. 2014;**32**:51603. DOI: 10.1116/1.4893428

[31] Maydannik PS, Plyushch A, Sillanpää M, Cameron DC, Cameron DC. Spatial atomic layer deposition: Performance of low temperature H_2O and O_3 oxidant chemistry for flexible electronics encapsulation. Journal of Vacuum Science and Technology A. 2015;**33**:031603. DOI: 10.1116/1.4914079

[32] Levy DH, Nelson SF, Freeman D. Oxide electronics by spatial atomic layer deposition. Journal of Display Technology. 2009;**5**:484-494. Available from: http://jdt.osa.org/abstract.cfm?URI=jdt-5-12-484 [Accessed: Jun 27, 2012]

[33] Nguyen VH, Resende J, Jiménez C, Deschanvres J, Carroy P, Muñoz D, et al. Deposition of ZnO based thin films by atmospheric pressure spatial atomic layer deposition for application in solar cells. Journal of Renewable and Sustainable Energy. 2017;**9**:021203. DOI: 10.1063/1.4979822

[34] Poodt P, Mameli A, Schulpen J, Roozeboom WMMEK, Roozeboom F. Effect of reactor pressure on the conformal coating inside porous substrates by atomic layer deposition. Journal of Vacuum Science and Technology A. 2017:021502. DOI: 10.1116/1.4973350

[35] Suh S, Park S, Lim H, Choi Y-J, Seong Hwang C, Joon Kim H, et al. Investigation on spatially separated atomic layer deposition by gas flow simulation and depositing Al_2O_3 films. Journal of Vacuum Science & Technology A—Vacuum Surfaces and Films. 2012;**30**:051504. DOI: 10.1116/1.4737123

[36] Ellinger CR, Nelson SF. Selective area spatial atomic layer deposition of ZnO, Al_2O_3, and aluminum-doped ZnO using poly(vinyl pyrrolidone). Chemistry of Materials. 2014;**26**: 1514-1522. DOI: 10.1021/cm402464z

[37] Ali K, Choi K-H, Muhammad NM. Roll-to-roll atmospheric atomic layer deposition of Al_2O_3 thin films on PET substrates. Chemical Vapor Deposition. 2014;**20**:1-8. DOI: 10.1002/cvde.201407126

[38] Poodt P, Knaapen R, Illiberi A, Roozeboom F, van Asten A. Low temperature and roll-to-roll spatial atomic layer deposition for flexible electronics. Journal of Vacuum Science & Technology A—Vacuum Surfaces and Films. 2012;**30**:01A142. DOI: 10.1116/1.3667113

[39] Levy DH, Ellinger CR, Nelson SF. Metal-oxide thin-film transistors patterned by printing. Applied Physics Letters. 2013;**103**:043505. DOI: 10.1063/1.4816322

[40] Lahtinen K, Maydannik P, Johansson P, Kääriäinen T, Cameron DC, Kuusipalo J. Utilisation of continuous atomic layer deposition process for barrier enhancement of extrusion-coated paper. Surface and Coatings Technology. 2011;**205**:3916-3922. DOI: 10.1016/j.surfcoat.2011.02.009

[41] Poodt P, van Lieshout J, Illiberi A, Knaapen R, Roozeboom F, van Asten A. On the kinetics of spatial atomic layer deposition. Journal of Vacuum Science & Technology A—Vacuum Surfaces and Films. 2013;**31**:01A108. DOI: 10.1116/1.4756692

[42] Poodt P, Illiberi A, Roozeboom F. The kinetics of low-temperature spatial atomic layer deposition of aluminum oxide. Thin Solid Films. 2013;**532**:22-25. DOI: 10.1016/j.tsf.2012.10.109

[43] Mousa MBM, Ovental JS, Brozena AH, Oldham CJ, Parsons GN. Modeling and experimental demonstration of high-throughput flow-through spatial atomic layer deposition of Al_2O_3 coatings on textiles at atmospheric pressure. Journal of Vacuum Science and Technology A. 2018;**36**:031517. DOI: 10.1116/1.5022077

[44] Celle C, Cabos A, Fontecave T, Laguitton B, Benayad A, Guettaz L, et al. Oxidation of copper nanowire based transparent electrodes in ambient conditions and their stabilization by encapsulation: Application to transparent film heaters. Nanotechnology. 2018;**29**:085701

[45] Yersak AS, Sharma K, Wallas JM, Dameron AA, Li X, Yang Y, et al. Spatial atomic layer deposition for coating flexible porous Li-ion battery electrodes Spatial atomic layer deposition for coating flexible porous Li-ion battery electrodes. Journal of Vacuum Science and Technology A. 2018;**36**:01A123. DOI: 10.1116/1.5006670

[46] Musselman KP, Muñoz-Rojas D, Hoye RLZ, Sun H, Sahonta S-L, Croft E, et al. Rapid open-air deposition of uniform, nanoscale, functional coatings on nanorod arrays. Nanoscale Horizons. 2017;**2**:110-117. DOI: 10.1039/C6NH00197A

[47] Franke S, Baumkötter M, Monka C, Raabe S, Caspary R, Johannes H-H, et al. Alumina films as gas barrier layers grown by spatial atomic layer deposition with trimethylaluminum and different oxygen sources. Journal of Vacuum Science & Technology A—Vacuum Surfaces and Films. 2017;**35**:01B117. DOI: 10.1116/1.4971173

[48] Ali K, Ali J, Mehdi SM, Choi K-H, An YJ. Rapid fabrication of Al2O3 encapsulations for organic electronic devices. Applied Surface Science. 2015;**353**:1186-1194. DOI: 10.1016/j.apsusc.2015.07.032

[49] Choi H, Shin S, Jeon H, Choi Y, Kim J, Kim S, et al. Fast spatial atomic layer deposition of Al_2O_3 at low temperature (<100 °C) as a gas permeation barrier for flexible organic light-emitting diode displays. Journal of Vacuum Science & Technology A—Vacuum Surfaces and Films. 2016;**34**:01A121. DOI: 10.1116/1.4934752

[50] Ali K, Choi K-H. Low-temperature roll-to-roll atmospheric atomic layer deposition of Al_2O_3 thin films. Langmuir. 2014;**30**:14195-14203. DOI: 10.1021/la503406v

[51] Illiberi A, Scherpenborg R, Roozeboom F, Poodt P. Atmospheric spatial atomic layer deposition of In-doped ZnO. ECS Journal of Solid State Science and Technology. 2014;**3**:P111-P114. DOI: 10.1149/2.002405jss

[52] Hoye RLZ, Muñoz-Rojas D, Iza DC, Musselman KP, MacManus-Driscoll JL. High performance inverted bulk heterojunction solar cells by incorporation of dense, thin ZnO layers made using atmospheric atomic layer deposition. Solar Energy Materials & Solar Cells. 2013;**116**:197-202. DOI: 10.1016/j.solmat.2013.04.020

[53] Illiberi A, Scherpenborg R, Theelen M, Poodt P, Roozeboom F. On the environmental stability of ZnO thin films by spatial atomic layer deposition. Journal of Vacuum Science & Technology A—Vacuum Surfaces and Films. 2013;**31**:061504. DOI: 10.1116/1.4816354

[54] Hoye RLZ, Chua MR, Musselman KP, Li G, Lai M-L, Tan Z-K, et al. Enhanced performance in fluorene-free organometal halide perovskite light-emitting diodes using tunable, low electron affinity oxide electron injectors. Advanced Materials. 2015;**20**: 1414-1419. DOI: 10.1002/adma.201405044

[55] Ehrler B, Musselman KP, Böhm ML, Morgenstern FSF, Vaynzof Y, Walker BJ, et al. Preventing interfacial recombination in colloidal quantum dot solar cells by doping the metal oxide. ACS Nano. 2013;**7**:4210-4220. DOI: 10.1021/nn400656n

[56] Illiberi A, Scherpenborg R, Wu Y, Roozeboom F, Poodt P. Spatial atmospheric atomic layer deposition of $Al_xZn_{1-x}O$. ACS Applied Materials & Interfaces. 2013;**5**:13124-13128. DOI: 10.1021/am404137e

[57] Illiberi A, Roozeboom F, Poodt P. Spatial atomic layer deposition of zinc oxide thin films. ACS Applied Materials & Interfaces. 2012;**4**:268-272. DOI: 10.1021/am2013097

[58] Frijters CH, Poodt P, Illiberi A. Atmospheric spatial atomic layer deposition of Zn(O,S) buffer layer for $Cu(In,Ga)Se_2$ solar cells. Solar Energy Materials & Solar Cells. 2016;**155**:356-361. DOI: 10.1016/j.solmat.2016.06.016

[59] Nelson SF, Levy DH, Tutt LW, Burberry M. Cycle time effects on growth and transistor characteristics of spatial atomic layer deposition of zinc oxide. Journal of Vacuum Science & Technology A—Vacuum Surfaces and Films. 2012;**30**:01A154. DOI: 10.1116/1.3670878

[60] Nguyen VH, Gottlieb U, Valla A, Muñoz D, Bellet D, Muñoz-Rojas D. Electron tunneling through grain boundaries in transparent conductive oxides and implications for electrical conductivity: The case of ZnO:Al thin films. Materials Horizons. 2018;**5**:715-726. DOI: 10.1039/C8MH00402A

[61] Khan A, Nguyen VH, Muñoz-Rojas D, Aghazadehchors S, Jiménez C, Nguyen ND, et al. Stability enhancement of silver nanowire networks with conformal ZnO coatings deposited by atmospheric pressure spatial atomic layer deposition. ACS Applied Materials & Interfaces. 2018;**10**:19208-19217. DOI: 10.1021/acsami.8b03079

[62] Najafi M, Zardetto V, Zhang D, Koushik D, Dorenkampe M, Creatore M, et al. Highly efficient semitransparent p-i-n planar perovskite solar cells by atmospheric pressure spatial atomic layer deposited ZnO. Solar RRL. 2018;**2**:1800147. DOI: 10.1002/solr.201800147

[63] Dunlop L, Kursumovic A, MacManus-Driscoll JL. Reproducible growth of p-type ZnO:N using a modified atomic layer deposition process combined with dark annealing. Applied Physics Letters. 2008;**93**:172111. DOI: 10.1063/1.3000604

[64] Hoye RLZ, Ehrler B, Böhm ML, Muñoz-Rojas D, Altamimi RM, Alyamani AY, et al. Improved open-circuit voltage in ZnO-PbSe quantum dot solar cells by understanding and reducing losses arising from the ZnO conduction band tail. Advanced Energy Materials. 2014;**4**:1301544. DOI: 10.1002/aenm.201301544

[65] Nandakumar N, Dielissen B, Garcia-Alonso D, Liu Z, Gortzen R, Kessels WMME, et al. Resistive intrinsic ZnO films deposited by ultrafast spatial ALD for PV applications. IEEE Journal of Photovoltaics. 2015;**5**:1462-1469. DOI: 10.1109/JPHOTOV.2015.2438644

[66] Hoye RLZ, Muñoz-Rojas D, Nelson SF, Illiberi A, Poodt P, Roozeboom F, et al. Research update: Atmospheric pressure spatial atomic layer deposition of ZnO thin films: Reactors, doping, and devices. APL Materials. 2015;**3**:040701. DOI: 10.1063/1.4916525

[67] Musselman KP, Albert-Seifried S, Hoye RLZ, Sadhanala A, Muñoz-Rojas D, MacManus-Driscoll JL, et al. Improved exciton dissociation at semiconducting polymer:ZnO donor: acceptor interfaces via nitrogen doping of ZnO. Advanced Functional Materials. 2014;**24**: 3562-3570 DOI: 10.1002/adfm.201303994

[68] Hoffmann L, Theirich D, Schlamm D, Hasselmann T, Pack S, Brinkmann KO, et al. Atmospheric pressure plasma enhanced spatial atomic layer deposition of SnOx as conductive gas diffusion barrier. Journal of Vacuum Science and Technology A. 2018;**36**:01A112. DOI: 10.1116/1.5006781

[69] Hoffmann L, Brinkmann KO, Malerczyk J, Rogalla D, Becker T, Theirich D, et al. Spatial atmospheric pressure atomic layer deposition of tin oxide as an impermeable Electron extraction layer for perovskite solar cells with enhanced thermal stability. ACS Applied Materials & Interfaces. 2018;**10**:6006-6013. DOI: 10.1021/acsami.7b17701

[70] Aghaee M, Maydannik PS, Johansson P, Kuusipalo J, Creatore M, Homola T, et al. Low temperature temporal and spatial atomic layer deposition of TiO_2 films. Journal of Vacuum Science & Technology A—Vacuum Surfaces and Films. 2015;**33**:041512. DOI: 10.1116/1.4922588

[71] Armstrong CL, Price MB, Muñoz-Rojas D, Davis NJKL, Abdi-Jalebi M, Friend RH, et al. Influence of an inorganic interlayer on exciton separation in hybrid solar cells. ACS Nano. 2015;**9**:11863-11871. DOI: 10.1021/acsnano.5b05934

[72] Muñoz-Rojas D, Sun H, Iza DC, Weickert J, Chen L, Wang H, et al. High-speed atmospheric atomic layer deposition of ultra thin amorphous TiO_2 blocking layers at 100 °C for inverted bulk heterojunction solar cells. Progress in Photovoltaics: Research and Applications. 2013;**21**:393-400. DOI: 10.1002/pip.2380

[73] Marin AT, Muñoz-Rojas D, Iza DC, Gershon T, Musselman KP, MacManus-Driscoll JL. Novel atmospheric growth technique to improve both light absorption and charge collection in ZnO/Cu_2O thin film solar cells. Advanced Functional Materials. 2013;**23**:3413-3419. DOI: 10.1002/adfm.201203243

[74] Muñoz-Rojas D, Jordan M, Yeoh C, Marin AT, Kursumovic A, Dunlop L, et al. Growth of 5 $cm^2 V^{-1} s^{-1}$ mobility, p-type Copper(I)oxide (Cu_2O) films by fast atmospheric atomic layer deposition (AALD) at 225 °C and below. AIP Advances. 2012;**2**:042179. DOI: 10.1063/1.4771681

[75] Fullager DB, Boreman GD, Ellinger CD, Hofmann T. Broadband optical properties of aluminium zinc oxide thin films prepared by spatial atomic layer deposition. Thin Solid Films. 2018;**653**:267-273. DOI: 10.1016/j.tsf.2018.03.047

[76] Hoye RLZ, Musselman KP, Chua MR, Sadhanala A, Raninga RD, Friend RH, et al. Bright and efficient blue polymer light emitting. Journal of Materials Chemistry C. 2015;**3**: 9327-9336. DOI: 10.1039/C5TC01581B

[77] Illiberi A, Frijters C, Ruth M, Bremaud D, Poodt P, Roozeboom F, et al. Atmospheric spatial atomic layer deposition of ZnOS buffer layers for flexible Cu(In,Ga)Se_2 solar cells. Journal of Vacuum Science and Technology A. 2018;**36**:051511. DOI: 10.1116/1.5040457

[78] Moitzheim S, Balder JE, Poodt P, Unnikrishnan S, De Gendt S, Vereecken PM. Chlorine doping of amorphous TiO_2 for increased capacity and faster Li^+-ion storage. Chemistry of Materials. 2017;**29**:10007-10018. DOI: 10.1021/acs.chemmater.7b03478

[79] Illiberi A, Cobb B, Sharma A, Grehl T, Brongersma H, Roozeboom F, et al. Spatial atmospheric atomic layer deposition of $In_xGa_yZn_zO$ for thin films transistors. ACS Applied Materials & Interfaces. 2015;**7**:3671-3675. DOI: 10.1021/am508071y

[80] Illiberi A, Grob F, Frijters C, Poodt P, Ramachandra R, Winands H, et al. High rate (~7 nm/s), atmospheric pressure deposition of ZnO front electrode for Cu(In,Ga)Se_2 thin-film solar cells with efficiency beyond 15%. Progress in Photovoltaics: Research and Applications. 2013;**21**:1559-1566. DOI: 10.1002/pip.2423

[81] Higgs DJ, Dumont JW, Sharma K, George SM. Spatial molecular layer deposition of polyamide thin films on flexible polymer substrates using a rotating cylinder reactor. Journal of Vacuum Science & Technology A—Vacuum Surfaces and Films. 2018;**36**:01A117

[82] Ellinger CR, Nelson SF. Design freedom in multilayer thin-film devices. ACS Applied Materials & Interfaces. 2015;**7**:4675-4684. DOI: 10.1021/am508088p

[83] Nelson SF, Ellinger CR, Levy DH. Improving yield and performance in ZnO thin-film transistors made using selective area deposition. ACS Applied Materials & Interfaces. 2015;**7**:2754-2759. DOI: 10.1021/am5077638

[84] Illiberi A, Poodt P, Bolt P-J, Roozeboom F. Recent advances in atmospheric vapor-phase deposition of transparent and conductive zinc oxide. Chemical Vapor Deposition. 2014; **20**:234-242. DOI: 10.1002/cvde.201400056

[85] Illiberi A, Scherpenborg R, Poodt P, Roozeboom F. (Invited) spatial atomic layer deposition of transparent conductive oxides. ECS Transactions. 2013;**58**:105-110. DOI: 10.1149/05810.0105ecst

[86] Sharma K, Routkevitch D, Varaksa N, George SM. Spatial atomic layer deposition on flexible porous substrates: ZnO on anodic aluminum oxide films and Al_2O_3 on Li ion battery electrodes. Journal of Vacuum Science & Technology A—Vacuum Surfaces and Films. 2016;**34**:01A146. DOI: 10.1116/1.4937728

[87] Bellet D, Lagrange M, Sannicolo T, Aghazadehchors S, Nguyen VH, Langley DP, et al. Transparent electrodes based on silver nanowire networks: From physical considerations towards device integration. Materials (Basel). 2017;**10**:570. DOI: 10.3390/ma10060570

[88] Langley D, Giusti G, Mayousse C, Celle C, Bellet D, Simonato J-P. Flexible transparent conductive materials based on silver nanowire networks: A review. Nanotechnology. 2013;**24**:452001. DOI: 10.1088/0957-4484/24/45/452001

[89] Sannicolo T, Lagrange M, Cabos A, Celle C, Simonato J, Bellet D. Metallic nanowire-based transparent electrodes for next generation flexible devices : A review. Small. 2016;**12**:6052-6075. DOI: 10.1002/smll.201602581

[90] Gregorczyk K, Knez M. Progress in materials science hybrid nanomaterials through molecular and atomic layer deposition: Top down, bottom up , and in-between approaches to new materials. Progress in Materials Science. 2016;**75**:1-37. DOI: 10.1016/j.pmatsci.2015.06.004

[91] Zhao B, Lee LC, Yang L, Pearson AJ, Lu H, She X, et al. In-situ atmospheric deposition of ultra-smooth nickel oxide for efficient perovskite solar cells. ACS Applied Materials & Interfaces. 2018. DOI: 10.1021/acsami.8b15503

[92] Creyghton Y, Illiberi A, Mione A, Boekel W. van, Debernardi N, Seitz M, et al. Plasma-enhanced atmospheric-pressure spatial ALD of Al_2 O_3 and ZrO_2Y. Creyghton. ECS Transation. 2016;**75**:11-19

[93] Mione MA, Katsouras I, Creyghton Y, van Boekel W, Maas J, Gelinck G, et al. Atmospheric pressure plasma enhanced spatial ALD of ZrO_2 for low-temperature, large-area applications. ECS Journal of Solid State Science and Technology. 2017;**6**:N243-N249

[94] Gregory G, Wilson M, Ali H, Davis KO. Thermally Stable Molybdenum Oxide Hole-Selective Contacts Deposited using Spatial Atomic Layer Deposition. 2018 IEEE 7th World Conference Photovolt. Energy Convers. (WCPEC)(A Jt. Conf. 45th IEEE PVSC, 28th PVSEC 34th EU PVSEC) (pp. 2006-2009). IEEE. 2006-2009 (2018). DOI: 10.1109/PVSC.2018.8547343

Chapter 2

Direct Liquid Injection Chemical Vapor Deposition

Vincent Astié, Cyril Millon,
Jean-Manuel Decams and Ausrine Bartasyte

Additional information is available at the end of the chapter

http://dx.doi.org/10.5772/intechopen.80244

Abstract

Thin film technology, based on different chemical and physical methods, enabled miniaturization, co-integration, and amelioration of the performance of the devices. Chemical vapor deposition (CVD) systems ensure high productivity and demonstrate excellent film uniformity (up to 12 inch wafers) and repeatability with high throughput for a variety of different films of oxides, nitrides, metals, chalcogenides, etc. In the last two decades, direct liquid injection (DLI)-CVD enabling the usage of solid and liquid precursors has proven to be one of the most versatile CVD process to meet industrial requirements. In this chapter, the requirements to the precursors suitable for DLI-CVD, different classes of available precursors, and models used to describe the evaporation are overviewed. Then, different liquid delivery devices used in DLI-CVD such as capillary tubes, syringes, aerosol delivery systems, and valves are reviewed in detail.

Keywords: direct liquid injection, DLI-CVD, evaporation, precursors, injectors, thin films

1. Introduction

In the last decades, thin film technology has been integrated with many large area applications in electronics, information processing/storage, telecommunications, LED lighting/displays, solar energy harvesting, etc. Thin film technology enabled miniaturization, co-integration, and amelioration of the performance of the devices. Many different chemical and physical deposition methods were developed in order to satisfy industrial needs and requirements in different application fields. There is no universal technique for the processing of functional coatings. For each application, more than one technique is being considered at industrial level and the final choice depends on the specifications and on the maturity and readiness level of each one. The industry prefers conservative approaches in terms of available subcontracting environment, which makes that process modification requires much higher performance, durability, and lower cost than actually implemented solutions, the safety/environmental issues are considered, as well.

There are two basic technologies, used for the synthesis of functional coatings: dry methods (depositions in vacuum conditions) and wet chemistry. The liquid-phase techniques, such as ink-jet printing, aerosol/spray deposition, sol-gel method, spin-coating, or dip-coating, are widely used in the coating technology due to their scalability and low cost [1, 2]. The usage of liquids/solvents in the wet techniques may induce degradation of the properties of the bottom layers and substrates and may introduce the impurities/defects in the heterostructure, which may be a critical limitation to device performance in the applications such as electronics [2]. Moreover, the drying of the solvent can also lead to the inhomogeneous distribution of the solute—"coffee stain" effect [3]. It is very challenging to grow the homogeneous coatings on 3D substrates with high aspect features by means of wet techniques due to dewetting and surface tension issues on the sidewalls and edges [2]. There is a considerable effort to be made to solve the strong drawbacks and limitations to the types of substrates of the wet techniques. Moreover, these methods usually provide poor reproducibility and non-uniform compositional distributions in the coatings. Thus, dry methods are highly preferred for the controlled synthesis of functional coatings. However, due to the strong advantages (low cost and easy up-scaling), wet techniques are able to meet actual specifications and thus take lead in "technology competition" with regard to dry techniques.

Different physical and chemical vapor deposition methods are used for the dry synthesis of the functional films at laboratory scale [4]. Physical vapor deposition techniques, such as pulsed laser deposition (PLD), magnetron sputtering, etc. offer in most cases a direct link between the composition of the gas phase and that of the coating [5], they may require less tuning and for this reason they are able to provide to the academia and the industry with "real samples" easier and faster than chemical vapor deposition techniques. However, these techniques will face issues in the stoichiometry adjustment in the case of multicomponent films if one of the deposited components is highly volatile (or presents very different evaporation characteristics). For instance, molecular beam epitaxy (MBE) is very powerful and essential tool in the semiconductor industry. Sputtering and evaporation techniques are widely applied for deposition of metal, oxide, and nitride films in the microfabrication of devices. Pulsed laser deposition (PLD) is broadly used by researchers for fundamental studies of new materials. However, physical methods, such as PLD, evaporation, and magnetron sputtering, are not compatible for depositions on non-planar surfaces and cannot offer precision control of the film thicknesses as required by some industries.

The industrial requirements (quality and reproducibility) for designed coatings and especially on 3D surfaces are more likely to be attained by chemical deposition methods, such as (metal organic) chemical vapor deposition (MOCVD) and atomic layer deposition (ALD) [6, 7]. ALD provides a tool for conformal coating on very high aspect-ratio nanostructures with excellent uniformity. It has become a technique for both template-directed nanofabrications and engineering of surface properties. For example, ALD is used to deposit HfO_2 layers with nanometer scale thickness in the complementary metal oxide semi-conductor (CMOS) fabrication. CVD systems ensured high productivity and demonstrated excellent film uniformity (up to 12 inch wafers) and repeatability with high throughput for a variety of different films including semiconductors (Si, III–V, etc.), dielectrics (e.g. SiO_2, AlN, etc.), rare-earth oxides, ceramic materials (e.g. TiN, ZrO_2, etc.), ferroelectrics (e.g. $PbTiO_3$, $LaNiO_3$, etc.), superconductors (e.g. $YBa_2Cu_3O_{7-x}$), chalcogenides, and noble metals. Despite industrial success of CVD in the manufacture of thin films and its flexibility, offered by chemistry, CVD is still an expensive method compared to the wet chemistry techniques.

In the CVD process, gaseous or evaporated precursors are transported to the heated substrate where decompositions or other type of reactions take place. However, not all precursors are under gaseous form at atmospheric pressure and ambient temperature (see **Figure 1**). In fact, only about 10% of the industrially available precursors are gaseous (e.g., trimethylborane $B(CH_3)_3$, trimethylsilane $Si(CH_3)_3$, etc.). The vast majority of precursors are available in solid form (powders or crystals) (80%). About 45% of elements can be grown by using liquid precursors and the precursors of 5% elements are only liquid. Most available precursors exhibit safety hazards and must be carefully operated and stored. The main disadvantages of classical MOCVD systems are the difficulty to control the chemistry and the evaporation of the solid/liquid precursors and therefore the composition and structure of coating (especially those of multicomponents coatings), which affect the reproducibility. It makes difficult the control of the film thickness and growth rate, which affects crystalline orientation, surface morphology, and film density. Nevertheless, chemistry of deposition process might be a potential advantage for precision manufacturing of films, if used in a controlled manner.

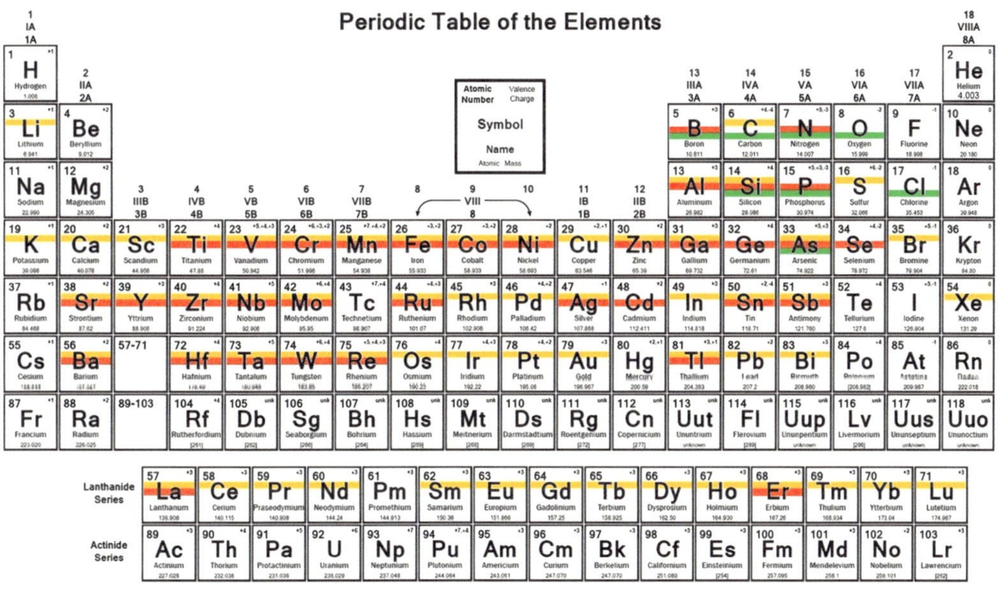

Figure 1. Physical states of commercially available precursors at ambient temperature for each element in the periodic table (gas in green, liquid in red, and solid in gold). www.sciencenotes.org

In the case of solid precursors, the sublimation rate is directly linked to its free surface, which in turn depends on the powder grain size and the amount of remaining precursor. Consequently, as sublimation occurs, its surface area in contact with the gas changes, thus changing the mass transport rate during the deposition. Due to the difficulty in precisely controlling the evaporation of a powder, depositions involving multiple solid precursors frequently were done by grinding optimized amounts of multiple precursors together to have a homogeneous single-source powder and rarely multiple evaporators were used. Various single-source powder precursors and techniques used to introduce powder in the evaporator were reviewed in Ref. [8]. In 1998, Gorbenko and Bosak [9] used a vibrating feeder to control the introduction of a mixture of β-diketonates precursors (see Section 2.1), namely $La(thd)_3$ and $Ni(acim)_2$ or $Ni(thd)_3$, for the deposition of $LaNiO_3$ thin films and succeeded in growing oriented stoichiometric layers. However, as many precursor powders are moisture sensitive, another method was later proposed in which pellets were covered by an inert material [10]. An alternative way was a packaging of a mixture of precursor powders in a quartz tube, which was then slowly introduced into the vaporizer. The large thermal gradient in the evaporator combined with the slow entrainment of the tube ensured that the growth rate would only be dictated by the density of the powder [11]. Although it worked well in laboratories, it proved to be difficult a scaling-up to industrial requirements.

The term "direct liquid injection (DLI)-CVD" refers to reactors, which use liquid delivery units to feed the deposition zone in reactants. It is much easier to regulate precisely a liquid flow rate than that of powders. In some systems, the liquid goes through an evaporation unit where it is vaporized, and in other cases, the precursors reach the surface of a heated substrate in a liquid state and then they are vaporized prior to decomposition. If the precursor is still in liquid phase when it reaches the substrate, the deposition method is called spray or aerosol pyrolysis [12] depending on the generation method of the liquid droplets. Sometimes spray/aerosol pyrolysis is called atmospheric pressure CVD as usually the evaporation takes place in proximity of the substrate heated to high temperatures. The main advantage of DLI-CVD is that, in most cases, the solution is kept under pressure at room temperature, which makes it possible to use precursors with low thermal stability and low vapor pressure. Moreover, powder precursors may be dissolved in the solvent and used in DLI-CVD systems, as well.

A bubbler technology is well established as a liquid precursor delivery system since many years [13]. The principle is rather simple and robust: the liquid is stored inside a so-called bubbler, typically a stainless steel canister, and an inert carrier gas (usually Ar or N_2) is introduced inside the liquid to bubble. The precursor vapor saturates the atmosphere in the bubbler and the vapor is then entrained to the heated substrate surface. The delivery of reactants depends on three parameters: the temperature of the bubbler, the carrier gas flow rate, and the pressure over the surface of the liquid. This system works particularly well with liquid precursors which vapor pressures are not too low and not too high. In the case of precursors with low vapor pressure, the only solution is heating the bubbler to increase the equilibrium vapor pressure inside the canister. Yet, this has some major drawbacks as it restricts the usage to precursors that are stable at high temperature, and implies to heat the whole delivery line so that condensation does not occur before reaching the substrate. To overcome the problems of non-saturation of the carrier gas and fluctuation of the precursor flow rate, many different devices were imagined, from two bubblers in series [14] to a self-metering reservoir [15]. Before 1990, the main research effort on the heretofore mentioned interesting materials were dedicated to physical deposition processes such as physical vapor deposition (PVD), as problems with vapor phase control and stoichiometry were not totally overcome in CVD.

In the 1990s, the growing interest in the thin films of superconducting materials, ferroelectrics, and high-κ dielectrics ($YBa_2Cu_3O_{7-x}$, $PbZr_{1-x}Ti_xO_3$, $BaTiO_3$, $SrTiO_3$, etc.) [16] leads to the development of more reliable and versatile delivery systems adapted to a wider range of precursors. The first generation of the IIA group solid precursors were thermally unstable at their sublimation temperature. They had a strong tendency to oligomerize and aging issues [17]. Much effort was made in precursor chemistry and in development of new liquid injection systems. In this chapter, we overview liquid delivery systems developed to date emphasizing on the problems it solved and the challenges it rose.

2. Precursors for DLI-CVD

2.1. Choice of the precursors

The injection devices have greatly enhanced the vaporization efficiency, thus the high volatility is not the main issue as it was with previous delivery systems. From a point of view for large-scale CVD applications, choosing the best precursors for a given application can sometimes be troublesome. As a matter of fact, a suitable precursor should meet several key requirements:

- **i.** High purity and high yield,
- **ii.** Stability at ambient temperature,
- **iii.** Easiness to handle and non-toxicity, and
- **iv.** Congruent volatilization with a significant temperature gap between evaporation and decomposition.

Additionally, it was shown that the molecular structure of the precursor had a significant impact on the growth dynamics and properties of the films, as well [18–20]. An *organometallic compound* should have a direct metal-carbon bond in their molecular structure, and it is admitted that precursors with a metal ion and a coordination complex (e.g. a metal ion is linked to a carbon chain through oxygen or nitrogen) are *metalorganic compounds*. Most of commercially available products belong to these five major classes of metalorganic compounds: metal β-diketonates, metal cyclopentadienyls, metal alkoxides, metal alkyls, and metal alkylamides depending on the organic complex associated to the metal.

Metal β-diketonates are the most used and common precursors for MOCVD and especially for depositions of oxide films. The (O,O′-η^2) chelating ligands of β-diketone usually form both thermally and hydrolytically stable complex with metal ion. Although it is mostly an advantage, it implies that the deposition temperature should be high enough to reach the precursor decomposition temperature or that additional source of energy (plasma, UV light, etc.) is provided at low deposition temperatures. However, it can heavily increase the carbon contamination when working at low temperatures. Additionally, their low reactivity with water has proven to be a complication in ALD processes. Many derivatives of β-diketonates precursors and the abbreviations of their names are presented in **Figure 2**. The volatility of the precursor can be tuned by modifying R-groups of ligands. For instance, Schweitzer et al. have shown that the lead precursors with bulkier R-groups became more volatile ($Pb(acac)_2 < Pb(dhd)_2 < Pb(thd)_2$) [21]. They explained this tendency by the increased shielding effect on the highly positive metal center and the reduced intermolecular oxygen-metal interactions in the solid state. Additionally, Fahlman et al. have reported that the increased fluorine substitution in the R-groups could improve significantly the volatility of zirconium β-diketonate precursors [22]. But on the other hand, fluorinated precursors often led to fluorine contamination in the deposited oxide layers. This contamination could be reduced by the introduction of excess H_2O vapor [23]. Another way to increase the volatility and/or the solubility in a solvent of the β-diketonate complex [24] is to add stoichiometric amount of a Lewis-base compound in the solution or directly on the complex in order to saturate the coordination sphere of the metal. The β-diketonates precursors have been successfully used to grow a large variety of films from pure metals, simple oxides, solid solutions, and composite materials (e.g., ruthenium and ruthenium oxide (RuO_2) [25], lanthanum-modified lead zirconate-titanate (PLZT) [26], yttria-stabilized zirconia (YSZ) [27], lithium niobate [27, 28], etc.).

Metal alkoxide ($M(OR)_n$) precursors are the second type of the most used precursors in MOCVD processes. They are usually more volatile than β-diketonates and provide less carbon contamination even at low oxygen partial pressure. Their volatility is linked to the electron density on the metallic center and the saturation level by ligands as it will condition the tendency to form oligomeric $(M(OR)_n)_m$ clusters [29]. The oligomerization can however

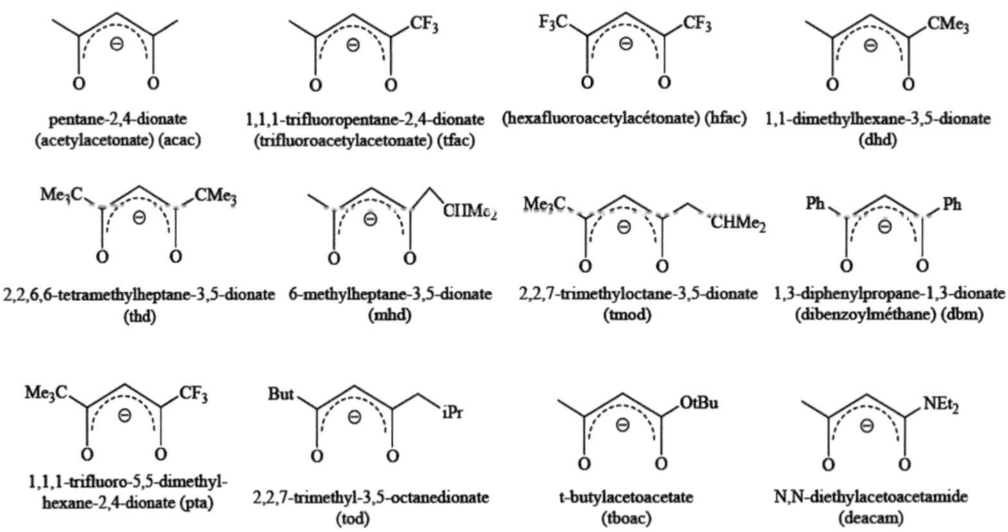

Figure 2. Composition, names/abbreviations of β-diketonate ligands [31].

be attenuated, or even suppressed, by using bulky ligand (e.g. isopropoxide (O^iPr) or tert-butoxide (O^tBu)) [30]. For example, the volatility of zirconium precursor increases in the order $Zr(OEt)_4 < Zr(O^iPr)_4 < Zr(O^tBu)_4$, the first being a trimer, the second a dimer, and the third a monomer. Another possible way is to use the alkoxides functionalized with a donor like dimethylaminoethoxide ($[OCH_2CH_2NMe_2]$ often referred to as (dmae)). However, metal alkoxide precursors can be hydrolyzed very easily [31] and some of them are highly toxic. These precursors too have been extensively used for the last two decades in the growth of oxide films like $LiNbO_3$ [32, 33], $LaAlO_3$ [34], YSZ [35], etc.

Metal cyclopentadienyl (Cp) precursors show good thermal stability and volatility. However, their η^5-bonding mode makes them a considerable source of carbon contamination. They are reactive with water even at moderate temperature and this makes them suitable candidates for ALD processes [36, 37]. Metal alkyl precursors are usually very reactive, toxic, and/or pyrophoric, as well as being a source of heavy carbon contamination in oxide films [38]. Nevertheless, aluminum and zinc alkyls are successfully used in ALD processes to grow Al_2O_3 from trimethylaluminum ($Al_2(CH_3)_6$) [39] and ZnO from diethylzinc ($Zn(C_2H_5)_2$ or DEZ) [40]. On the other hand, metal alkylamides usually have medium volatility and some of them exhibit good stability to hydrolysis. Most of alkylamides can be easily aminolyzed. Thus, they are reliable precursor sources for the growth of metal nitrides. Although they do not have any metal-oxygen bond, high purity oxide films were grown under big oxygen partial pressures by MOCVD [41] and ALD [42].

As briefly presented, different types of metalorganic precursors have their own strengths and weaknesses and the operators should know the deposition temperature before choosing their precursors. Most of β-diketonates will show good yield at high growth temperatures and they will be more resistant to hydrolysis but a synthesis will be inefficient at low temperatures. The alkoxides could be more adapted to depositions at low temperature where the main issue is the carbon contamination. The cyclopentadienyls have shown good efficiency in ALD processes. To overcome limitations of different families of precursors, a number of scientific teams have tailored new precursors combining the good stability of β-diketonates with the high volatility of alkoxides and have shown that the precursors consisting of different types of ligands showed better performance than those with single type of ligands. In fact, Jones et al. [19] showed that $Zr_2(O^iPr)_6(thd)_2$ was more suitable precursor for the growth of $Pb(Zr,Ti)O_3$ at moderate temperatures than the commercially available $Zr(thd)_4$. The influence of precursor structure on GaAs, ZnSe, and $Al_xGa_{1-x}N$ film growth was described in detail by Jensen et al. [18].

2.2. Evaporation of liquid droplets

The theory describing the heat transfer between a gaseous medium and a droplet of liquid is well known and the basics of the evaporation process were given by Maxwell in 1877 [43]. His model for stationary state evaporation [44] assumed that:

i. The droplet was spherical and without any relative motion with regards to the gas;

ii. The vapor concentration at the surface of the droplet was equal to the saturation concentration; and

iii. The evaporation was an equilibrium process.

However, in the case of droplet evaporation with a low vapor pressure, this model is not sufficient. There are six different models, used to explain vaporizing process at low vapor pressure:

i. The constant-droplet-temperature model (decreasing linearly as a function of the square of the droplet diameter, d, also called d^2 law);

ii. The infinite-liquid-conductivity model;

iii. The spherically symmetric transient droplet-heating (or conduction-limit) model;

iv. The effective-conductivity model;

v. The vortex model; and

vi. The Navier-Stokes equation solution.

The infinite-liquid-conductivity model is commonly used in modeling of industrial spray processes. In order to simplify the description, this model is based on several assumptions:

i. A single Lewis number in the gas phase surrounding the droplet and

ii. A negligible transient liquid heating although it would appear to be a governing factor of the droplet vaporization rate [45].

Detailed description of fluid dynamics and transport of droplets and sprays can be found in Ref. [46]. This model was used to describe the fuel droplet vaporization in combustion processes [47, 48]. Although the description of evaporation phenomenon is complex and many assumptions are needed, basically it can be easily understood, that smaller droplets are more easily evaporated. In a heated environment at low vacuum, the heat is transferred through the exchange between the gas and the droplet. Two parameters have to be taken into account: the specific heat capacity, $C_{p,}$ of the gas and the free surface of the droplet. The surface/volume ratio can be increased by minimizing the radius and evaporation at a given flow rate can be ameliorated by introducing more but smaller droplets. This possibility was not really considered in the first liquid delivery devices, but very quickly the atomization has become one of the major concerns of the injector industry and the key parameter, used to discriminate liquid delivery devices.

3. Liquid delivery systems

In this section, different liquid delivery devices used in DLI-CVD such as capillary tubes, syringes, aerosol delivery systems, and valves and their developments are reviewed in detail.

3.1. Capillary tubes and syringes

Capillary tubes are one of the simplest ways to introduce a flow of precursor with controlled rate into an evaporator. The liquid drops into a heated zone (either a quartz tube [49] or a stainless steel "tee" [50]) with flowing a neutral gas, or more conventionally, a nebulizer. The principle of this method is based on the capillary effect. The flow inside the capillary is laminar, the liquid in contact with the walls is stationary and there is no radial flow. Thus, the pressure across the cross-section is constant. According to Hagen-Poiseuille law, the flow volume through the tube, Q, is governed by four parameters such as the viscosity of the precursor solution, μ; the radius of tube, r; the length of the cylindrical tube, d; and lastly the pressure difference between the top (usually sealing to the bottle which contains the precursor) and the bottom "pressure drop", ΔP:

$$Q = \frac{\pi r^4}{8\mu} * \frac{\Delta P}{d} \tag{1}$$

The main advantage of this liquid delivery system is an ability to provide a steady flow of precursors to the vaporizer. However, the substantial drawback is that the flow volume itself can only be tuned by varying the viscosity of the liquid. This means changing the solvent or the concentration of the precursor in the solution, which in turn could affect the quality of the deposition. Borgharkar et al. [50] have even reported that increasing the concentration of $Cu(hfac)_2$, diluted in isopropanol (i-PrOH), led to an unstable flow rate through the capillary due to high viscosity.

To tune the injected flow itself, syringes were used as the precursor vessel, and the process was sometimes referred to as lirect liquid evaporation-CVD (DLE-CVD) [51]. The syringe is placed into a syringe driver, or syringe pump, connected to the capillary tube. The liquid flow is controlled precisely by adjusting the rotational speed of a stepper motor. The multiple precursors can be introduced to the evaporator by means of either multiple syringes and their drivers, or by using their mixture in a single solution. However, this technology is well adapted only to the liquid precursors at ambient temperature and the usage of the precursor solutions may face several issues. Indeed, since the bottom of the capillary tube is in contact with a heated zone, the vapor pressure difference between the diluted solid precursor and the solvent may lead to the evaporation of the solvent with low vapor pressure prior to exiting the capillary. This in turn can introduce a partial/complete clogging of the tip with the precipitated solid precursor and disruption of the solution flow. The usage of heavy solvents may eliminate these issues but it can have a significant impact on the carbon contamination of the deposited layer. A possible solution to this problem was proposed by ASM International by an introduction a three-way valve in order to introduce a carrier gas together with the solution in the capillary system [16]. An example of such system using an advanced syringe-capillary system to control the precursor injection rate, adapted to the synthesis of carbon nanotubes, is presented in **Figure 3** [52]. Many different thin films of metals and binary oxides thin were grown using this technique: Cu [50], transparent amorphous Al_2O_3 [53], ZrO_2 [54], TiO_2 [55], La_2O_3 [56], VO_2 [57], etc.

3.2. Aerosol assisted-CVD and pulsed-pressure MOCVD

Some CVD processes rely on the nebulization, or the atomization of liquid precursors into a "mist", or an aerosol, which is then brought to the heated substrate by carrier gases. The ultrasonic atomization process takes roots in the Lord Rayleigh's "*The Theory of Sound*" [58], and Lang's works on the absorption of energy by a liquid [59]. When a liquid is placed on a flat surface vibrating along the surface normal, some of the energy is absorbed and transformed

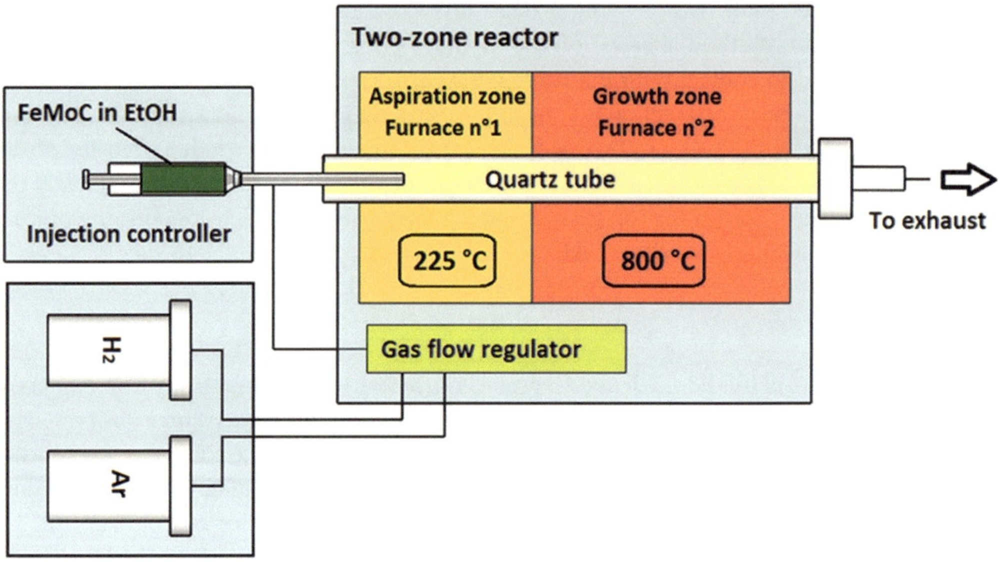

Figure 3. Schematic representation of a DLE-CVD reactor using a syringe pump associated with a 9″ needle. Reproduced with permission [52]. Copyright 2018, MDPI.

into transverse standing waves called capillary waves in the liquid. As the amplitude of the vibration increases, the difference in height between the crests and troughs increases to the point where it compromises the stability of the wave. At this stage, the waves collapse and tiny droplets of liquid are ejected from the crests. If the vibrational frequency is too high, a cavitation can occur and large portions of liquid are sent [60].

Size of droplets, produced by industrial atomizers, follows a narrow Gaussian distribution and the mean size value depends entirely on the wavelength of the capillary wave, which in turn depends on the frequency of the transducer and the properties of the liquid [58, 59]:

$$\lambda_L = \sqrt[3]{\left(\frac{8\pi^*\gamma}{\rho^* f^2}\right)} \text{ and} \tag{2}$$

$$D_{N,0.5} = 0{,}34^*\lambda_L \tag{3}$$

where λ_L is a wavelength of the liquid, γ is a surface tension, ρ is the density, f is a frequency of the actuator, and $D_{N,0.5}$ is the median diameter of droplets. Tsai et al. have demonstrated as expected that the drop size could be decreased by increasing the ultrasonic frequency [61]. However, it should be noted that the working frequency of some atomizers cannot be tuned. In fact, the working frequency of an ultrasonic spray nozzle is governed by its dimensions. The high-frequency nozzles work with small wavelength and are consequently smaller than low-frequency nozzles. The rate of liquid flow to atomizer can be calibrated and controlled accurately by means of liquid mass flow controller. Theoretically, a very large range of flow rates should be possible, but in fact, the ratio of maximum to minimum flow rates produced by the same atomizer is limited to 5 due to design constraints [62]. Usually, atomizer works with a piezoelectric transducer, which converts an electrical input into a mechanical output. This transformation is accompanied by a heating of both the transducer and the precursor solution resulting in the shift in the resonance frequency and in the change of the liquid viscosity, respectively. This alters the flow rate of the liquid resulting in an unstable film growth. Thus, most deposition cycles to finish in a relatively short time (matter of minutes) [63]. Ultrasonic atomization in aerosol-assisted CVD (AACVD) has the benefit to allow a wider choice of precursors. In fact, a high volatility is not anymore a requirement and a solubility in solvent becomes the main selection criteria [64].

Different CVD systems based on aerosol delivery systems have been developed with customized vaporization zones. In the case of AACVD, also referred to as liquid source misted chemical deposition (LSMCD) [63] or liquid source chemical vapor deposition (LSCVD) [65, 66], the precursors or their solution are sent continuously to the reactor where they evaporate in the vicinity of the substrate surface due to high thermal gradient. Although McMillan et al. [67] have reported that in the case of LSCVD, an uniform film deposition could not be achieved on a rotating heated substrate unless a barrier plate was located in the vicinity of its surface, the homogeneous depositions were demonstrated by later developed AACVD systems. A possibility to grow thin films of a large variety of oxides and metals was studied by using an ultrasonic delivery systems. As illustrative examples of AACVD could be:

i. The deposition of partially reduced tungsten oxide, WO_{3-x}, from polyoxotungstate anions $[n\text{-}Bu_4N]_2[W_6O_{19}]$ and $[n\text{-}Bu_4N]_4H_3[PW_{11}O_{39}]$ using a PIFCO ultrasonic humidifier [64] and

ii. The growth of nanocomposite of VO_2 thin films with embedded cerium dioxide CeO_2 and titanium dioxide TiO_2 nanoparticles by means of a hybrid method combining the atmospheric pressure (AP)-AACVD using standard Vicks humidifier with a continuous hydrothermal flow synthesis (CHFS) [68].

In the case of pulsed-pressure CVD (PP-CVD) [69] or pulsed liquid injection with an ultrasonic nozzle [70], a small shot of liquid is pushed by an inert gas into the vaporizer through an ultrasonic nozzle (**Figure 4**), and quickly evaporates [71]. This produces a pressure pulse, and then the chamber is subsequently pumped back to the initial pressure in order to send another shot. High efficiency of deposition process has been achieved by means of PP-CVD [72]. Conventional nebulizer was employed in PP-MOCVD such as Sono-Tek ultrasonic nozzle. PP-MOCVD was used to grow sealing layers of yttria-stabilized zirconia on porous solid oxide fuel cell electrodes [35], tantalum oxide [73], epitaxial titanium dioxide [71, 74, 75], as well as lithium tantalate [76].

3.3. Valves

The liquid delivery systems in CVD, presented previously, were developed by implementing different industrially available valves. According to the operational mode, valves can be classified into ON/OFF valves and proportional valves. The proportional valve allows very precise control of the opening area via a piezoelectric material, which in return allows for an adjustable flow rate. The ON/OFF operating valve was typically used in the motor industry prior to the 2014 Euro 6 emission standards. In the 1980s, electromagnetic ON/OFF valves, in which a current is applied to the solenoid generating attraction force in order to pull the plunger below up, releasing the valve seat and letting the liquid flow out of the chamber through a very fine nozzle [77], are well known [78, 79] and used in many ON/OFF systems where the control of

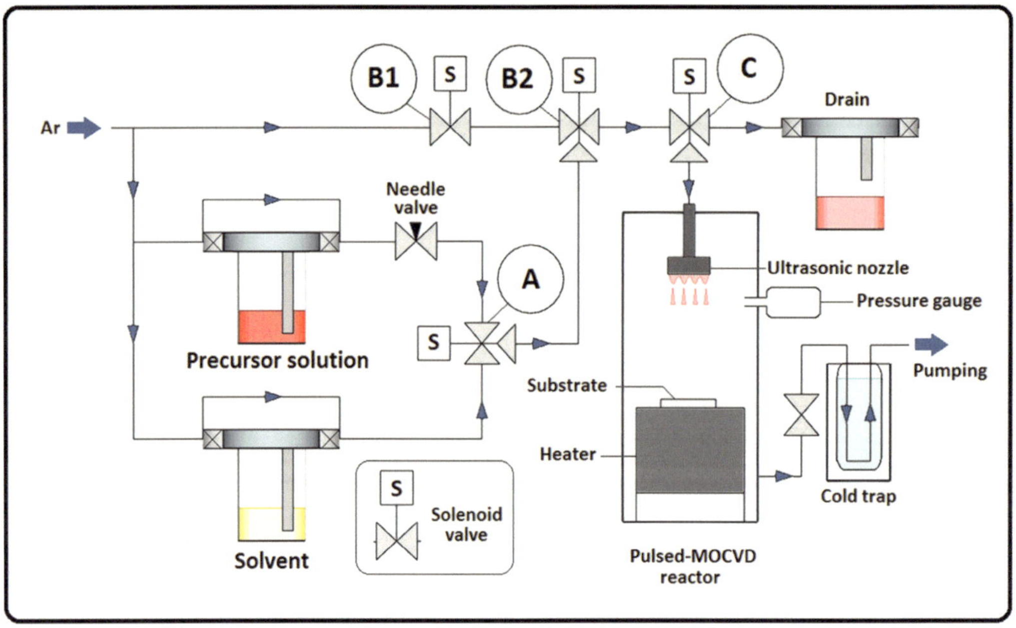

Figure 4. Schematic representation of a PP-CVD reactor using an ultrasonic nozzle. Reproduced with permission [75]. Copyright 2001, Elsevier.

an injected volume is crucial. In 1993, Sénateur et al. have proposed to use them in a pulsed injection (PI)-MOCVD process [80]. A strong advantage of PI-MOCVD systems is a clear interface between the pressurized environment at ambient temperature where the precursor is stored safely and the low-vacuum atmosphere of the heated evaporator. Thus, the problems related to vapor pressure differences between the solvent and its solute could be mostly overlooked in the case of PI-MOCVD. The injector works as a valve, regulating the flow of liquid introduced in the evaporation chamber and as a nozzle atomizing the flow into tiny droplets.

As explained earlier (see Section 2.2), the dimensions of droplets have a fundamental impact on the evaporation efficiency of the liquid solution. Therefore, there are several key parameters of conventional solenoid injectors defining a quality of atomization: (i) the nozzle geometry (size, number of orifices, and degree of conicity), (ii) the flow regime inside the injector, and (iii) properties of the solution/liquid such as viscosity and concentration. The nozzle geometries and flow regimes have been extensively studied in the fuel injection industry [81–85]. Most commercially available injectors work in the frequency range between 0.1 and 200 Hz with very small opening times. These characteristics offer a very large working range in terms of injected flow rate and ultimately, in terms of film growth rate. In such injection system, the injection flow rate depends on the concentration and viscosity of the precursor solution, the opening frequency and time of the nozzle, and the pressure difference between the stored solution and the evaporator. Very small opening time ensures the fast evaporation process and helps to avoid the dissociation or premature decomposition of precursors before evaporation. However, it should be noted that the injector may produce instable flow at opening time below 1 ms. It is important to note, that injectors are commanded by computer, which allows digital control of the injection frequency, number of injected droplets, opening time, and consequently thin film growth rate and thickness. As in the case of ultrasonic nozzles, there are usually no corrosion problems inside the injector with organic precursors/solvents, but some applications may require specially coated injectors.

It was shown that the precursors with low vapor pressure can be efficiently used in PI-MOCVD. PI-MOCVD reactors equipped with multiple injectors are used for the growth of multi-element films with controlled composition/stoichiometry, multilayers, nanocomposites and complex nanostructures of superconductors [16], high-κ dielectrics [86, 87], ferroelectrics, conductors [88], etc.

As the solvent can cause a carbon contamination in the films, Kaul and Seleznev [89] have proposed a method to remove the solvent of a solution prior to vaporizing the precursor. In this process, a fiberglass belt is wetted with a controlled amount of solution outside of the evaporation chamber. Then, the tape is brought into a warm zone of the reactor where only the solvent evaporates. The belt with solid precursor is then mechanically carried to a zone where flash evaporation of precursors occurs. Later, the same process has been applied with in situ wetting of the tape with sequential injection of micro-amounts of solution (illustrated in **Figure 5**) [90]. Although a solvent was introduced like in PI-MOCVD, only a mixture of precursor vapor and carrier gas reaches the substrate. This method may be particularly useful in the synthesis at low temperatures in order to solve carbon contamination issues.

An injection system was first commercialized under the JetPulse® name by Jipelec, a division of Qualiflow, and it was later acquired and further developed by AIXTRON® under the name TriJet® [91]. TriJet® evaporators are installed in the latest AIXTRON® ALD machines and have been used in MOCVD processes, called atomic vapor deposition (AVD). In fact, the reproducibility of the commercially available injectors was very poor. To face fluctuations in the partial pressure of the liquid, operators had either to dilute significantly the precursor solution or to lower the injection frequency. Additionally, since the nozzle of the injector is subjected to the heated environment, a premature evaporation of liquid solution trapped inside the injector may happen, which can significantly alter the injection flow rate of diluted

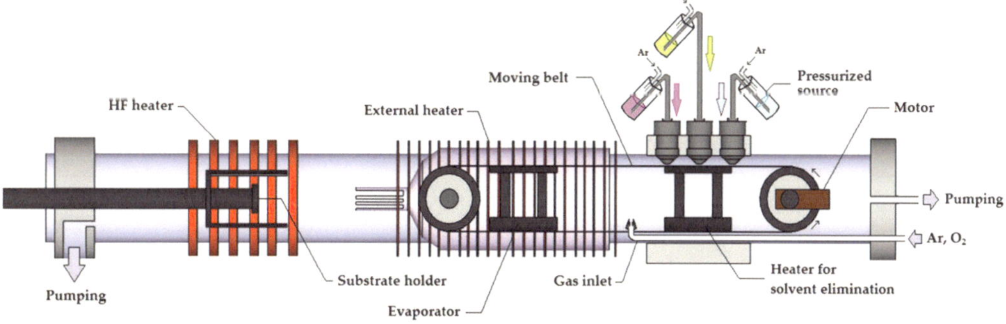

Figure 5. Schematic representation of a DLI-CVD furnace using solenoid electrovalves in combination with a moving tape. Reproduced with permission [90]. Copyright 1998, Elsevier.

solids and clog partially/completely the injector [92]. To overcome these issues, Kemstream® came up with the idea to have a two-stage injection process (**Figure 6**). The liquid injector introduces a precursor solution into the "mixing chamber", where a controlled amount of carrier gas is mixed with the precursor solution. A few milliseconds later, the second injector opens and the mixture is injected and atomized into the evaporator. The operator must make sure that there is a short delay between the opening times of two injectors. The most important advantage of double injector is the formation of a homogeneous liquid-gas phase in the mixing chamber. This drastically diminishes the radius of the injected droplets, hence increasing the evaporation efficiency. This method offers a far better atomization of the precursor solution along with the possibility to use even lower vapor pressure precursors than in single-injection systems. Brooks® Instrument also proposes a multiple liquid inlet direct liquid vaporizer system (**Figure 6**), which injects the solution into a chamber where it is both atomized by sending a carrier gas perpendicularly to the stream and vaporized in contact with a hot gas [93]. Kemstream®'s injector (**Figure 7a**) can be found on Annealsys® CVD and ALD machines and process are called "direct liquid injection (DLI)-CVD and ALD" [94–96]. The Brooks® DLI Vaporizer Systems (**Figure 7b**) were successfully implemented in industrial CVD [97–99] and ALD [34, 100–105] processes, as well. The major part of the DLI-CVD industry is dedicated to the deposition of very well known III–V semiconductors, in which HORIBASTEC Liquid Vaporization Systems (**Figure 7c**) are broadly used [106]. It is also interesting to note that although the latest emission standards pushed the fuel industry to go toward piezoelectric injectors, the piezoelectrically driven injectors and piezo-valves are still very rare in DLI-CVD systems [107].

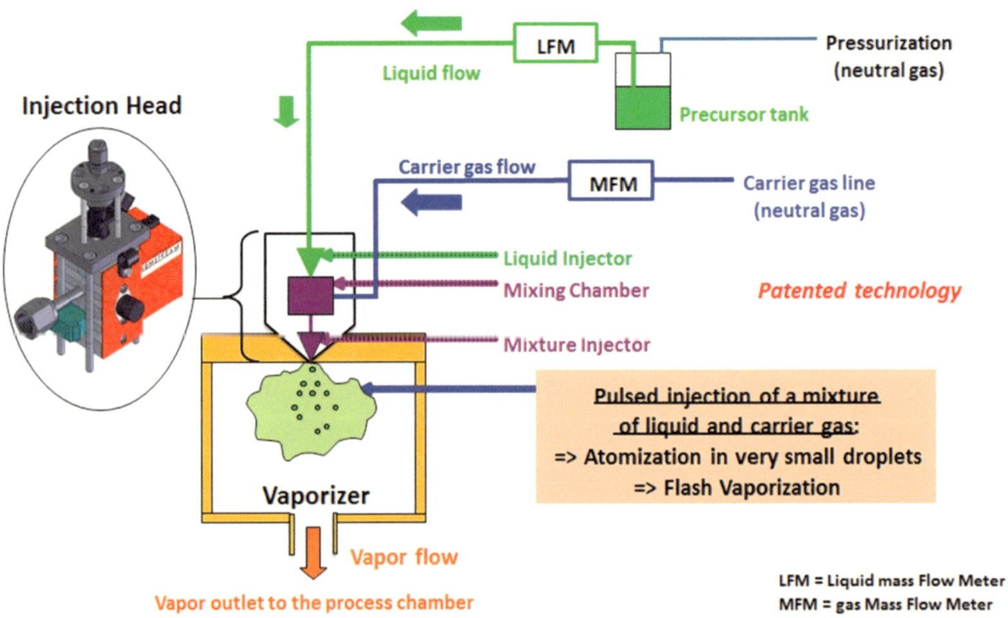

Figure 6. Schematic representation of Kemstream® two-stage injection process.

http://dx.doi.org/10.5772/intechopen.80244

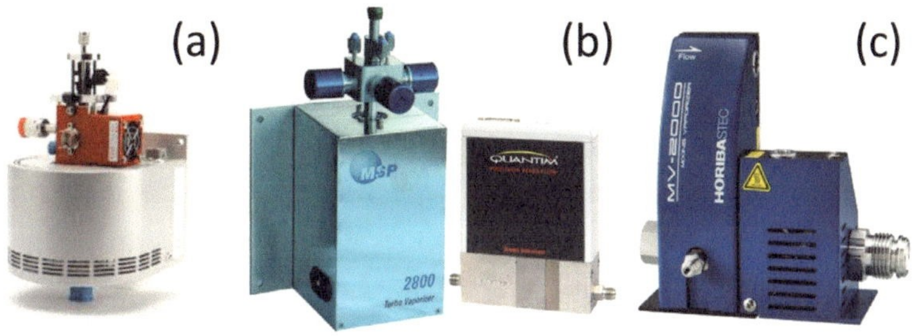

Figure 7. Kemstream DLI injector plugged into a vaporizing box (a), Brooks DLI vaporizer system (b), and HORIBASTEC liquid vaporization systems (c).

4. Conclusions and perspectives

Liquid mass flow controllers appear as a viable option for controlling a rate of a liquid flow, but they cannot be used as a direct liquid delivery unit in CVD systems. The main problem lies in the fact that it delivers a continuous flow of precursors in the evaporator. The process then depends entirely on the ability of the evaporator to vaporize the precursor, which is inefficient without any other atomizing system. The usage of a heated plate would be very unreliable in time. In the case of nebulizers or transducers, the progressive heating of the element leads to a drift in the aerosol flow. On the other hand, the ultrasonic nozzle shows a good reliability in evaporation and a good capability to handle "difficult" precursors. Each nozzle works at a certain frequency dictated by its dimension and it limits the tenability of the feeding rate. To regulate the amount of precursor sent to the substrate, the operator can either dilute/concentrate the solution or limit/increase the flow that comes to the nozzle.

Injection systems based on solenoid valves have reached very good performance and technological maturity. Major advantages brought by injection delivery systems are versatility, the digital control of the growth rate and the thickness of the film, and the storage of precursors at ambient conditions ensuring their stability in time. Advanced new-generation injection systems guarantee a controllable and stable feeding rate. DLI-CVD permits to vary and to control easily the composition of complex material film and doping level from sample to sample. Thus, it provides an excellent film composition control (<1%), has low cost, high efficiency, offers high uniformity on large area (<1.5%, up to 12 inch wafers), good conformal step coverage, high flexibility in precursors selection (solid or liquid), atomic control of film thickness (up to 10 Å), and possibility to deposit complex materials and multilayers. DLI-CVD is widely applied in industry of thin films fabrication ensuring high productivity and demonstrating excellent film uniformity and repeatability with high throughput in the CMOS era for a variety of different films including high-k dielectrics, rare-earth oxides, perovskite oxide and chalcogenide, and noble metals.

Although several solutions were proposed in the literature, the industrial DLI-CVD systems are still not well adapted to the depositions at low temperatures due to issues of carbon contamination originating from organic part of the precursors and the presence of the solvent. Thus, further developments continue on the low-temperature CVD systems assisted by additional energy source such as plasma, UV, etc. Many research teams work on the molecular engineering of the precursor structure including combination of multiple ligands in order to ameliorate the decomposition of precursors at low deposition temperatures. CVD system efficiency and application at industrial scale remains highly dependent on the available precursor quality and reliability.

As described above, direct liquid delivery CVD has taken a profit from the developments in other fields, for example, chromatography, industrial spray coating, and automotive industry. The valves, transducers, liquid mass flow controllers, capillaries, etc. developed for other applications were successfully adapted to the liquid precursor/solution delivery in the CVD reactors. Thus, the advances in liquid delivery systems mainly used in large public applications determine the further developments of CVD systems, as well. The automotive industry has upgraded its fuel injection technology by implementing injectors with piezoelectric nozzles to adapt to the new emission standards. The piezoelectric nozzles react much faster, meaning a faster opening and closure of the valve assuring a better control of the injected flow, and the opening time can be tuned much easier by changing the allocated tension. Probably, all these improvements will be soon implemented in the next generation DLI-CVD systems, which will serve beyond CMOS Era.

Acknowledgements

The authors are grateful to the French National Research Agency for funding through the grants ANR LiLit (ANR-16-CE24-0022-011), Labex ACTION program (ANR-11-LABX-0001-01), and the EUR EIPHI program (contract no. ANR-17-EURE-0002). This work was partly supported by the French RENATECH network. The authors are also grateful to A. Almirall, S. Kuprenaite, and G. Clementi for the fruitful discussions and corrections.

Author details

Vincent Astié[1]*, Cyril Millon[1], Jean-Manuel Decams[2] and Ausrine Bartasyte[1]

*Address all correspondence to: vincent.astie@femto-st.fr

1 FEMTO-ST Institute, University of Bourgogne Franche-Comté, CNRS UMR 6174, Besançon, France

2 Annealsys, Montpellier, France

References

[1] Braun D, Cherdron H, Rehahn M, Ritter H, Voit B. Polymer Synthesis: Theory and Practice: Fundamentals, Methods, Experiments. 5th ed. Heidelberg: Springer-Verlag Berlin; 2013

[2] Ozaydin-Ince G, Coclite AM, Gleason KK. CVD of polymeric thin films: Applications in sensors, biotechnology, microelectronics/organic electronics, microfluidics, MEMS, composites and membranes. Reports on Progress in Physics. 2012;**75**:016501

[3] Deegan RD, Bakajin O, Dupont TF, Huber G, Nagel SR, Witten TA. Capillary flow as the cause of ring stains from dried liquid drops. Nature. 1997;**389**:827

[4] Jensen KF, Kern W, Vossen JL. Thin Film Processes II. San Diego, California: Academic Press Inc.; 1991. p. 284

[5] Reichelt K, Jiang X. The preparation of thin films by physical vapour deposition methods. Thin Solid Films. 1990;**191**:91

[6] Choy KL. Chemical vapour deposition of coatings. Progress in Materials Science. 2003; **48**:57

[7] Liu M, Li X, Karuturi SK, Tokb AIY, Fan HJ. Atomic layer deposition for nanofabrication and interface engineering. Nanoscale. 2012;**4**:1522

[8] Choy KL. Innovative Processing of Films and Nanocrystalline Powders. Singapore: World Scientific; 2002

[9] Gorbenko OY, Bosak AA. Growth of $LaNiO_3$ thin films on MgO by flash MOCVD. Journal of Crystal Growth. 1998;**186**(1-2):181-188

[10] Chuprakov IS, Martin JD, Dahmen KH. New flash-evaporation feeder for chemical vapor deposition. Le Journal de Physique IV. 1999;**9**(PR8):Pr8-901-Pr8-908

[11] Hiskes R et al. Single source MOCVD of epitaxial oxide thin films. Journal of Crystal Growth. 1993;**128**:781-787

[12] Perednis D, Gauckler LJ. Thin film deposition using spray pyrolysis. Journal of Electroceramics. 2005;**14**:103-111

[13] Manasevit HM, Simpson WI. The use of metal-organics in the preparation of semiconductor materials I. Epitaxial gallium-V compounds. Journal of the Electrochemical Society. 1969;**116**(12):1725-1732

[14] Dickinson PH et al. Chemical vapor deposition of $YBa_2Cu_3O_{7-x}$ superconducting films. Journal of Applied Physics. 1989;**66**(1):444-447

[15] Rodgers DB. Self-metering Reservoir. Rhinebeck, New York: Icon Dynamics LLC; 1997

[16] Sénateur JP et al. Pulsed injection MOCVD of functional electronic oxides. Advanced Materials for Optics and Electronics. 2000;**10**:155-161

[17] Sievers RE et al. Volatile barium β-diketonates for use as MOCVD precursors. Coordinatio Chemistry Reviews. 1993;**128**:285-291

[18] Jensen KF et al. Metalorganic chemical vapor deposition: Examples of the influence of precursor structure on film properties. Le Journal de Physique IV. 1991;**02**(C2):C2-243-C2-252

[19] Jones AC et al. Metalorganic chemical vapour deposition (MOCVD) of zirconia and lead zirconate titanate using a novel zirconium precursor. Journal of the European Ceramic Society. 1999;**19**:1431-1434

[20] Loboda MJ. Low temperature PECVD growth and characterization of a-SiC: H films deposited from silacyclobutane and silane/methane precursor gases. In: Amorphous and Crystalline Silicon Carbide IV. Heidelberg: Springer, Berlin; 1992. pp. 271-280

[21] Schweitzer GK, Pullen BP, Fang Y-H. The volatilities of some lead β-diketonates. Analytica Chimica Acta. 1968;**43**:332-334

[22] Fahlman BD, Barron AR. Substituent effects on the volatility of metal β-diketonates. Advanced Materials for Optics and Electronics. 2000;**10**:223-232

[23] Watson IM. Metal-organic CVD of the high-Tc superconductor $YBa_2Cu_3O_{7-\delta}$. Chemical Vapor Deposition. 1997;**3**(1):9-26

[24] Jodin L et al. Method for producing lithium-based layers by cvd. 2013. Google Patents

[25] Lee D-J, Kang S-W, Rhee S-W. Chemical vapor deposition of ruthenium oxide thin films from Ru(tmhd)3 using direct liquid injection. Thin Solid Films. 2002;**413**:237-242

[26] Tao W, Desu SB, Li TK. Direct liquid injection MOCVD of high quality PLZT films. Materials Letters. 1995;**23**:177-180

[27] Matsuno S et al. $YBa_2Cu_3O_x$ thin films with yttria stabilized zirconia buffer layer on metal substrate by liquid source chemical vapor deposition using tetrahydrofuran solution of β-diketonates. Japanese Journal of Applied Physics. 1995;**34**:2293

[28] Bartasyte A et al. Toward high-quality epitaxial $LiNbO_3$ and $LiTaO_3$ thin films for acoustic and optical applications. Advanced Materials Interfaces. 2017;**4**(8):1600998

[29] Jones AC. Molecular design of improved precursors for the MOCVD of electroceramic oxides. Journal of Materials Chemistry. 2002;**12**(9):2576-2590

[30] Bradley DC, Mehrotra RC, Gaur DP. Alkoxides, Metal. New York: Academic Press; 1978

[31] El Eter M. Synthèse, caractérisation et évaluation de nouveaux précurseurs azotés pour dépôt de films d'oxydes métalliques MO_2 (M = Hf, Zr) par MOCVD à injection liquide. Lyon: Université Claude Bernard-Lyon I; 2008

[32] Akiyama Y et al. Epitaxial growth of lithium niobate film using metalorganic chemical vapor deposition. Thin Solid Films. 2007;**515**(12):4975-4979

[33] Tanaka A et al. Preparation of lithium niobate films by metalorganic chemical vapor deposition with a lithium alkoxide source. Journal of Crystal Growth. 1995;**148**:324-326

[34] Gaskell JM et al. Liquid injection ALD and MOCVD of lanthanum aluminate using a bimetallic alkoxide precursor. Journal of Materials Chemistry. 2006;**16**(39):3854-3860

[35] Krumdieck SP et al. Solid yttria-stabilized zirconia films by pulsed chemical vapor deposition from metal-organic precursors. Journal of the American Ceramic Society. 2002; **85**(11):2873-2875

[36] Niinistö J et al. Advanced cyclopentadienyl precursors for atomic layer deposition of ZrO_2 thin films. Journal of Materials Chemistry. 2008;**18**(28):3385-3390

[37] Niinistö J et al. Controlled growth of HfO_2 thin films by atomic layer deposition from cyclopentadienyl-type precursor and water. Journal of Materials Chemistry. 2005;**15**(23): 2271-2275

[38] Frenck H et al. Low temperature remote plasma-enhanced deposition of thin metal oxide films by decomposition of metal alkoxides. Materials Science and Engineering: A. 1991;**139**:394-400

[39] Goldstein DN, McCormick JA, George SM. Al_2O_3 atomic layer deposition with trimethylaluminum and ozone studied by in situ transmission FTIR spectroscopy and quadrupole mass spectrometry. The Journal of Physical Chemistry C. 2008;**112**:19530-11939

[40] Lim SJ, Kwon S, Kim H. ZnO thin films prepared by atomic layer deposition and rf sputtering as an active layer for thin film transistor. Thin Solid Films. 2008;**516**(7):1523-1528

[41] Williams PA et al. Growth of hafnium dioxide thin films by liquid-injection MOCVD using alkylamide and hydroxylamide precursors. Chemical Vapor Deposition. 2003;**9**(6): 309-314

[42] Hausmann DM et al. Atomic layer deposition of hafnium and zirconium oxides using metal amide precursors. Chemistry of Materials. 2002;**14**(10):4350-4358

[43] Maxwell JC. Collected Scientific Papers. Taken from Fuchs NA. Evaporation and Droplet Growth in Gaseous Media. Cambridge, New York: Pergamon Press; 1959;**11**:1877

[44] Duguid HA. A study of the evaporation rates of small freely falling water droplets. Master Theses 5295, Rolla: University of Missouri; 1969

[45] Sirignano WA. Fuel droplet vaporization and spray combustion theory. Progress in Energy and Combustion Science. 1983;**9**(4):291-322

[46] Sirignano WA. Fluid Dynamics and Transport of Droplets and Sprays. Irvine: Cambridge University Press, University of California; 1999

[47] Kuo KK. Principles of Combustion. New York: John Wiley and Sons; 1986

[48] Williams FA. Combustion Theory, Benjamin Cummins. CA: Menlo Park; 1985

[49] Sinnott SB et al. Model of carbon nanotube growth through chemical vapor deposition. Chemical Physics Letters. 1999;**315**:25-30

[50] Borgharkar NS et al. Solution delivery of Cu(hfac)2 for alcohol-assisted chemical vapor deposition of copper. Journal of the Electrochemical Society. 1999;**146**(3):1041-1045

[51] Li K. Direct Liquid Evaporation Chemical Vapor Deposition (DLE-CVD) of Nickel, Manganese and Copper-Based Thin Films for Interconnects in Three-Dimensional Microelectronic Systems. Cambridge: Harvard University; 2016. p. 127

[52] Esquenazi GL, Brinson B, Barron AR. Catalytic growth of carbon nanotubes by direct liquid injection CVD using the nanocluster $[H_xPMo_{12}O_{40}\subset H_4Mo_{72}Fe_{30}(O_2CMe)_{15}O_{254}(H_2O)_{98-y}(EtOH)_y]$. C-Journal of Carbon Research. 2018;**4**(1):17

[53] Gordon RG, Kramer K, Liu X. Chemical vapor deposition and properties of amorphous aluminum oxide films. MRS Proceedings. 1996;**446**:383

[54] Jones AC et al. MOCVD of zirconia thin films by direct liquid injection using a new class of zirconium precursor. Chemical Vapor Deposition. 1998;**4**(2):46-49

[55] Awaluddin A et al. Direct liquid injection MOCVD growth of TiO_2 films using the precursor Ti(mpd)(dmae)2. Le Journal de Physique IV. 2001;**11**(PR3):Pr3-531-Pr3-537

[56] Kang S-W, Rhee S-W. Deposition of La_2O_3 films by direct liquid injection metallorganic chemical vapor deposition. Journal of the Electrochemical Society. 2002;**149**(6):C345-C348

[57] Vernardou D, Pemble ME, Sheel DW. Vanadium oxides prepared by liquid injection MOCVD using vanadyl acetylacetonate. Surface and Coatings Technology. 2004;**188-189**: 250-254

[58] Rayleigh JWS. The Theory of Sound. Vol. 2. London: MacMillan; 1896

[59] Lang RJ. Ultrasonic atomization of liquids. The Journal of the Acoustical Society of America. 1962;**34**(1):6-8

[60] Berger HL. Ultrasonic Liquid Atomization: Theory and Application. Hyde Park, New York: Partridge Hill Publishers; 1998

[61] Tsai S, Childs P, Luu P. Correlation between droplet diameter and viscosity. AICHE Journal. 1996;**42**(120):3340

[62] Engie R. Maximizing the use of platinum catalyst by ultrasonic spray application. In: Proceedings of ASME 2011 5th International Conference on Energy Sustainability & 9th Fuell Cell Science, Engineering and Technology Conference; Wasington, DC, USA. 2011

[63] Huffman M. Liquid source misted chemical deposition (LSMCD)—A critical review. Integrated Ferroelectrics. 1995;**10**(1-4):39-53

[64] Cross WB, Parkin IP. Aerosol assisted chemical vapour deposition of tungsten oxide films from polyoxotungstate precursors: Active photocatalysts. Chemical Communications. 2003;**14**:1696-1697

[65] McMillan LD, Paz de Araujo CA. U.S. Patent number 5.119.760. 1992

[66] McMillan LD, Paz de Araujo CA. U.S. Patent number 5.138.520. 1992

[67] McMillan LD et al. Liquid source CVD. Integrated Ferroelectrics. 1991;**2**:351-359

[68] Warwick MEA et al. Hybrid chemical vapour and nanoceramic aerosol assisted deposition for multifunctional nanocomposite thin films. Thin Solid Films. 2011;**519**(18):5942-5948

[69] Krumdieck SP. Pulsed-pressure MOCVD science. Materials and Technology. ECS Transactions. 2009;**25**(8):1209-1219

[70] Versteeg VA, Avedisian CT, Raj R. Method and Apparatus for CVD using Liquid Delivery System with an Ultrasonic Nozzle. USA: Cornell Research Foundation, Inc.; 1995

[71] Krumdieck SP, Raj R. Experimental characterization and modeling of pulsed MOCVD with ultrasonic atomization of liquid precursor. Chemical Vapor Deposition. 2001;**7**(2):85-90

[72] Krumdieck SP, Raj R. Conversion efficiency of alkoxide precursor to oxide films grown by an ultrasonic-assisted, pulsed liquid injection, metalorganic chemical vapor deposition (pulsed-CVD) process. Journal of the American Ceramic Society. 1999;**82**(6):1605-1607

[73] Porporati A, Roitti S, Sbaizero O. Metallorganic chemical vapor deposition of Ta_2O_5 films. Journal of the European Ceramic Society. 2003;**23**(2):247-251

[74] Versteeg VA, Avedisian CT, Raj R. Metalorganic chemical vapor deposition by pulsed liquid injection using an ultrasonic nozzle: Titanium dioxide on sapphire from titanium(IV) isopropoxide. Journal of the American Ceramic Society. 1995;**78**(10):2763-2768

[75] Krumdieck S, Raj R. Growth rate and morphology for ceramic films by pulsed-MOCVD. Surface and Coatings Technology. 2001;**141**(1):7-14

[76] Xie H, Raj R. Epitaxial $LiTaO_3$ thin film by pulsed metalorganic chemical vapor deposition from a single precursor. Applied Physics Letters. 1993;**63**(23):3146-3148

[77] Bosch Premium Fuel Injectors Trade Brochure. Robert Bosch LLC, Farmington Hills, Michigan. 17032145 RevA. 2017

[78] Paule K. Fuel Injection System for Internal Combustion Engines with an Electromagnetic Injector. Stuttgart: Robert Bosch GmbH; 1957

[79] Kennedy HE, Brooks W. Method and apparatus for the electromagnetic control of the fuel injection valves of an internal combustion engine. Patent no. DE655355C licenced to BROOKS WALKER. 1930

[80] Sénateur JP et al. Procédé et Dispositif d'Introduction de Précurseur dans une Enceinte de Dépôt Chimique en Phase Vapeur. Patent no. 93/08838 licenced to J.I.P. Elec, 11 chemin du Vieux Chêne, 38240 Meylan, France. 1993

[81] Reitz RD. Atomization and Other Breakup Regimes of a Liquid Jet. Princeton, New Jersey: Princeton University; 1978

[82] Tennant PA et al. Simulation of flow through single orifice nozzles. In ICLASS-90, Hartford, Connecticut. 1990

[83] Hiroyasu H, Arai M, Shimizu M. Break-up length of a liquid jet and internal flow in a nozzle. In ICLASS-91, Gaithersburg. 1991:275-292

[84] Schmidt DP, Corradini ML. The internal flow of diesel fuel injector nozzles: A review. International Journal of Engine Research. 2005;**2**(1):1-22

[85] Payri R et al. Diesel nozzle geometry influence on spray liquid-phase fuel penetration in evaporative conditions. Fuel. 2008;**87**(7):1165-1176

[86] Dubourdieu C et al. Pulsed liquid-injection MOCVD of high-K oxides for advanced semiconductor technologies. Materials Science and Engineering: B. 2005;**118**(1-3):105-111

[87] Dubourdieu C et al. Growth by liquid-injection MOCVD and properties of HfO2 films for microelectronic applications. Chemical Vapor Deposition. 2006;**12**(2-3):187-192

[88] Hušeková K et al. Growth of RuO_2 thin films by liquid injection atomic layer deposition. Thin Solid Films. 2010;**518**(16):4701-4704

[89] Kaul AR, Seleznev BV. New principle of feeding for flash evaporation MOCVD devices. Journal De Physique. IV. 1993;**3**(3):375-378

[90] Pignard S et al. Chemical vapor deposition and structural study of $La_{0.8}MnO_{3-\delta}$ thin films. Thin Solid Films. 1998;**347**:161-166

[91] Senzaki Y et al. Atomic level solutions® for advanced microelectronic applications. In: Solid-State and Integrated-Circuit Technology, ICSICT 2008. IEEE. 2008:1308-1311

[92] Guillon H et al. Device for Introducing, Injecting or Spraying a Mixture of a Carrier Gas and Liquid Compounds and Method for Implementing Said Device. Montpellier: KEMSTREAM; 2016

[93] DS-VAP-DLI Systems-eng (0611) Data Sheet: Liquid Vaporizer Systems, Brooks Instrument, LLC, Hatfield, Pennsylvania. 2011

[94] Avril L, Decams JM, Imhoff L. Pulsed direct liquid injection ALD of TiO_2 films using titanium tetraisopropoxide precursor. Physics Procedia. 2013;**46**:33-39

[95] Avril L et al. Alumina particle reinforced TiO_2 composite films grown by direct liquid injection MOCVD. Vacuum. 2014;**107**:259-263

[96] Boisselier G, Maury F, Schuster F. SiC coatings grown by liquid injection chemical vapor deposition using single source metal-organic precursors. Surface and Coatings Technology. 2013;**215**:152-160

[97] Li N et al. Growth of atomically smooth epitaxial nickel ferrite films by direct liquid injection CVD. Chemical Vapor Deposition. 2011;**17**(7-9):261-269

[98] Archana PS et al. Photocatalytic water oxidation at bismuth vanadate thin film electrodes grown by direct liquid injection chemical vapor deposition method. International Journal of Hydrogen Energy. 2017;**42**(12):8475-8485

[99] Yengantiwar A et al. Direct liquid injection chemical vapor deposition of molybdenum-doped bismuth vanadate photoelectrodes for efficient solar water splitting. The Journal of Physical Chemistry C. 2017;**121**(11):5914-5924

[100] Gaskell JM et al. Deposition of lanthanum zirconium oxide high-k films by liquid injection ALD and MOCVD. Chemical Vapor Deposition. 2007;**13**(12):684-690

[101] O'Kane R et al. Growth of HfO_2 by liquid injection MOCVD and ALD using new hafnium-cyclopentadienyl precursors. Chemical Vapor Deposition. 2007;**13**(11):609-617

[102] Black K et al. Deposition of ZrO_2 and HfO_2 thin films by liquid injection MOCVD and ALD using ansa-metallocene zirconium and hafnium precursors. Journal of Materials Chemistry. 2008;**18**:4561-4571

[103] Chalker PR et al. Liquid injection atomic layer deposition of silver nanoparticles. Nanotechnology. 2010;**21**(40):405602

[104] Karim Z et al. Advances in ALD equipment for sub-40nm memory capacitor dielectrics: Precursor delivery. Materials and Processes. ECS Transactions. 2008;**16**(4):125-134

[105] Kim SK, Hoffmann-Eifert S, Waser R. Growth of noble metal Ru thin films by liquid injection atomic layer deposition. The Journal of Physical Chemistry C. 2009;**113**:11329-11335

[106] PGV-CE23A, Liquid Source Vaporization System Datasheet. Kyoto: Horiba Stec, Co., Ltd.; 2018

[107] Hemmer E et al. Nanostructured ZrO_2 membranes prepared by liquid-injection chemical vapor deposition. Microporous and Mesoporous Materials. 2012;**163**:229-236

Section 2

CVD of Thin Films

Chapter 3

DLC Layers Created Using CVD Techniques and Their Application

Marta Januś

Additional information is available at the end of the chapter

http://dx.doi.org/10.5772/intechopen.79526

Abstract

One way to obtain new materials with different properties is to modify existing ones to improve their inadequate properties. Due to the fact that many useful properties of materials, including resistance to wear and corrosion, coefficient of friction and biocompatibility, depend on the state of the surface, modern methods of surface engineering are particularly useful. They include the deposition of layers with a matching chemical composition and structure. In terms of applications, the most suitable seem to be amorphous or nanocrystalline layers containing carbon, nitrogen, silicon and hydrogen. They combine the advantageous properties of silicon carbide SiC and silicon nitride Si_3N_4 and thus have a strong resistance to oxidation at high temperatures, high modulus of elasticity, low coefficient of friction and wear resistance. However, the silicon carbonitride compound is not thermodynamically stable under normal conditions and therefore must be obtained as a result of unconventional synthesis. One of these methods is chemical vapor deposition (CVD), including the most widely used plasma-assisted chemical vapor deposition (PACVD) techniques. The materials obtained, thanks to these techniques, have found and are still finding wider and wider application in many branches of industry and medicine.

Keywords: CVD, plasma, DLC layers, polymers, Ti alloy, Al alloy

1. Introduction

For several dozen years, scientists have been involved in obtaining diamond-like carbon (DLC) layers by various methods. Depending on the predominance of phenomena or the type of physical, chemical, or physicochemical interaction on the core or the substrate, it can be divided into six groups of methods: mechanical, thermomechanical, thermal, electrochemical, chemical, and physical [1]. Among them, plasma-assisted chemical vapor deposition (PACVD) techniques are most commonly used. These methods, thanks to the activation of chemical processes in the gas phase, which is a low-temperature plasma, allow for obtaining layers at relatively low temperatures [2, 3].

The history of amorphous carbon nitride (a-C:N:H) layers originated in the 1990s of the last century, when Liu and Cohen [4], using numerical methods, suggested the stability of β-C_3N_4 crystal (**Figure 1**). According to them, this crystal would have a structure analogous to β-Si_3N_4, characterized by short bonds and high covalence (about 93%). According to the predictions of these authors, C_3N_4 should have a modulus of stiffness and hardness even greater than diamond. Attractive, predicted properties of this material have become the cause of intense research on the synthesis of this crystal. In most experiments, however, amorphous materials were obtained with maximum nitrogen consisting of up to 45%. Only in a small number of cases, evidence for the existence of C_3N_4 in crystalline form has been found. In works [5, 6], small crystals in an amorphous matrix were observed under a transmission electron microscope (TEM).

Failed attempts to synthesize crystalline C_3N_4 lead to an increased interest in amorphous carbon nitride in the form of layers. These layers, as it turned out, are characterized by high wear resistance and a low coefficient of friction and thus can become competitive to diamond-like carbon (DLC) layers, which are well known and used in various branches of technology and medicine. This fact makes the a-C:N:H layers the focus of many research centers in the world.

Pioneer, though unsuccessful, research in crystalline C_3N_4 production also resulted in the discovery of the new SiC_xN_y silicon carbonitride compound, which combines the properties of silicon nitride, silicon carbide, and carbon nitride. Depending on the synthesis conditions and the precursors used, this compound may exist in both crystalline and amorphous forms [8, 9], and its chemical composition may change.

For example, at low temperatures, using precursors containing hydrogen in their composition, amorphous layers are formed in which this element occurs in their structure in a bonded form (SiC_xN_y(H)) [10, 11].

It is understood that the properties of these layers depend on their chemical composition and structure.

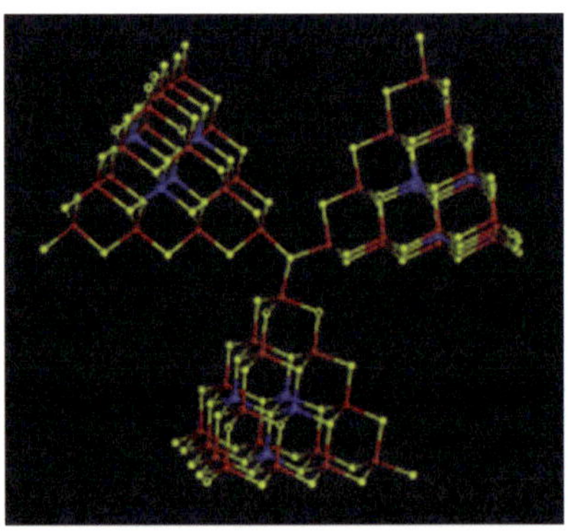

Figure 1. Computer model C3N4 [7].

http://dx.doi.org/10.5772/intechopen.79526

Materials with this type of applied layers have found wide application in many branches of the industry (electronics, optics, aviation, shipbuilding, construction) and in medicine.

A more detailed discussion of these problems will be provided in this chapter.

2. CVD techniques

In CVD processes, the growth of the coating is limited by the rate of chemical reactions between the adsorbed gas molecules on the surface of the substrate. At low temperatures, the growth of the layers is limited by chemical reactions and at high temperatures by diffusion of active products to the surface.

The partition of CVD methods (**Figure 2**) can be carried out in terms of:

- Gas substrate type: halide (H) and organometallic (MO)
- Pressure: atmospheric (AP) and depressed (LP)

The later methods (LP CVD) are divided by the way in which the process is activated:

- Plasma (PA):
  - Glow discharge (DC)
  - Radio frequency (RF)
  - Microwaves (MW)

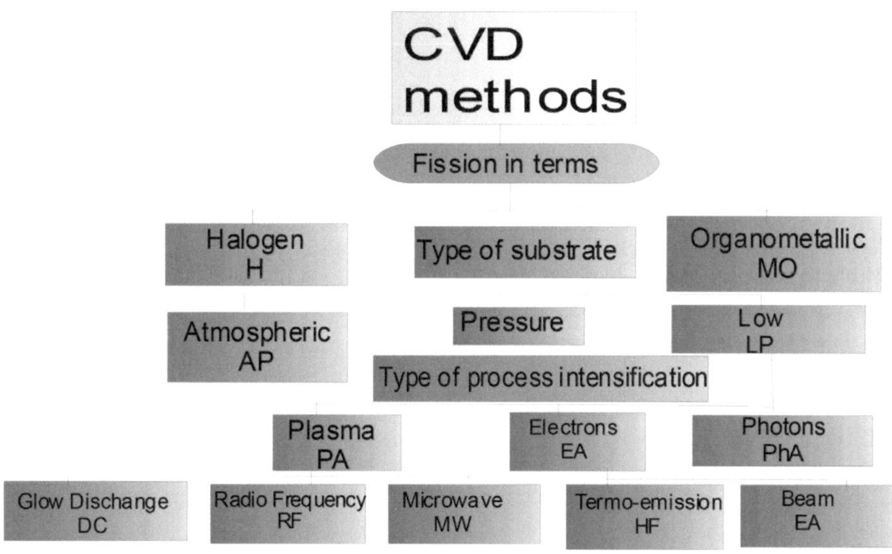

Figure 2. The partition of CVD methods [1].

- Electrons (EA)
    - Thermoemission (HF)
    - Bunch (EA)
- Photons (PhA)

The main difference between conventional and plasma-activated CVD methods lies in the fact that, in the case of PACVD, the molecules are in a higher state of internal energy. Electric discharges of 2.45 GHz (microwaves) and 13.56 MHz (radio waves) are among the most useful methods of producing low-temperature plasma [12, 13].

Figure 3a shows the reactions occurring in the CVD process involving plasma. The presence of particles with an increased state of internal energy and ion bombardment on the surface of the substrate increases the surface diffusion rate and decreases the activation energy of chemical reactions leading to the formation of a coating. Layering takes place at a much lower temperature.

This raises the question of what can affect the design of coating technologies. This can be considered from two sides:

- The need to meet the requirements in relation to the physicochemical parameters of the surface of the workpiece.
- Compliance with the economic aspect.

For processes taking place under reduced pressure, aside from the choice of the method, the individual processing phases are of particular importance. An important role is also played by the type of intermediate layers and final obtained coatings and thus the technological parameters of each stage.

Temperature is one of the process parameters. If CVD processes are generated by plasma, the treatment can already be carried out at room temperature. This is related to the activation of chemical reactions in the gas phase by electrons accelerated in an electromagnetic field.

In that case of coating deposition, with time, subsequent stages of layer growth occur, which consequently form a continuous structure on the surface (**Figure 3b**).

In addition to the substrate temperature and the power of the plasma generator, the technological parameters related to the gas precursors themselves, i.e., their type and partial pressure, also have an influence on the structure and properties of the obtained coatings. Controlling these parameters also affects the microstructure of the resulting layers [14], which can be:

- Amorphous
- Fine grained
- Polycrystalline

2.1. The role of plasma

The commonly known definition of plasma is, according to Langmuir [15], partially or almost completely ionized gas, which is also the fourth state of matter (97% of the universe [16]). Initially, it was an object of interest only among physicists. However, application in materials technology was also quickly found.

Research into the improvement of physical properties, as well as physicochemical properties, such as wear and corrosion resistance of thin films and coatings, has somehow contributed to the use of plasma energy for obtaining materials in the form of layers and coatings on various types of surface substrates.

The reactor chamber is a kind of unloading pipe in which (under reduced pressure (about 10–80 Pa), the influence of electromagnetic waves is present) gas ionizing energy is supplied with a strictly defined and controlled flow rate [17]. The results of this process are also:

- Reactions between dissociated molecules that may be neutral or ionized.
- Increasing the temperature of the substrate, ions and neutral molecules with high kinetic energy falling on its surface can be additionally supported by the use of heating elements in the system.
- Transport and adsorption processes of atoms and ions—under conditions of temperature gradient and concentration of components on the surface of the substrate—and their transport toward the core of the material.
- Coating growth—heterogeneous reactions in thermodynamically privileged places on the surface (active sites).

2.2. Synergism in coating technology

The concept of synergism is used in various fields of science. As a technique, it is usually referred to as the interaction of more than one factor for the final effect obtained (improvement of a given physicochemical property or obtaining a completely new functionality).

This definition is also reflected in surface technology. However, it is assumed here every time that the obtained effects are due to the influence of many factors [1]. In this case the term synergism refers here to the usable properties of the modified surface. This action is achieved by:

- Choosing the right method for treating the substrate surface.
- Selecting chemical and phase composition of coatings.
- Using transitional layers (so-called interlayers), this has a significant impact on adhesion of the obtained coatings to the surface.

Therefore, we can say that it is necessary to design technology in many experimental works that ensures optimal material parameters that are maximally beneficial for the given application (both the properties of the coatings obtained and their adhesion to the surface).

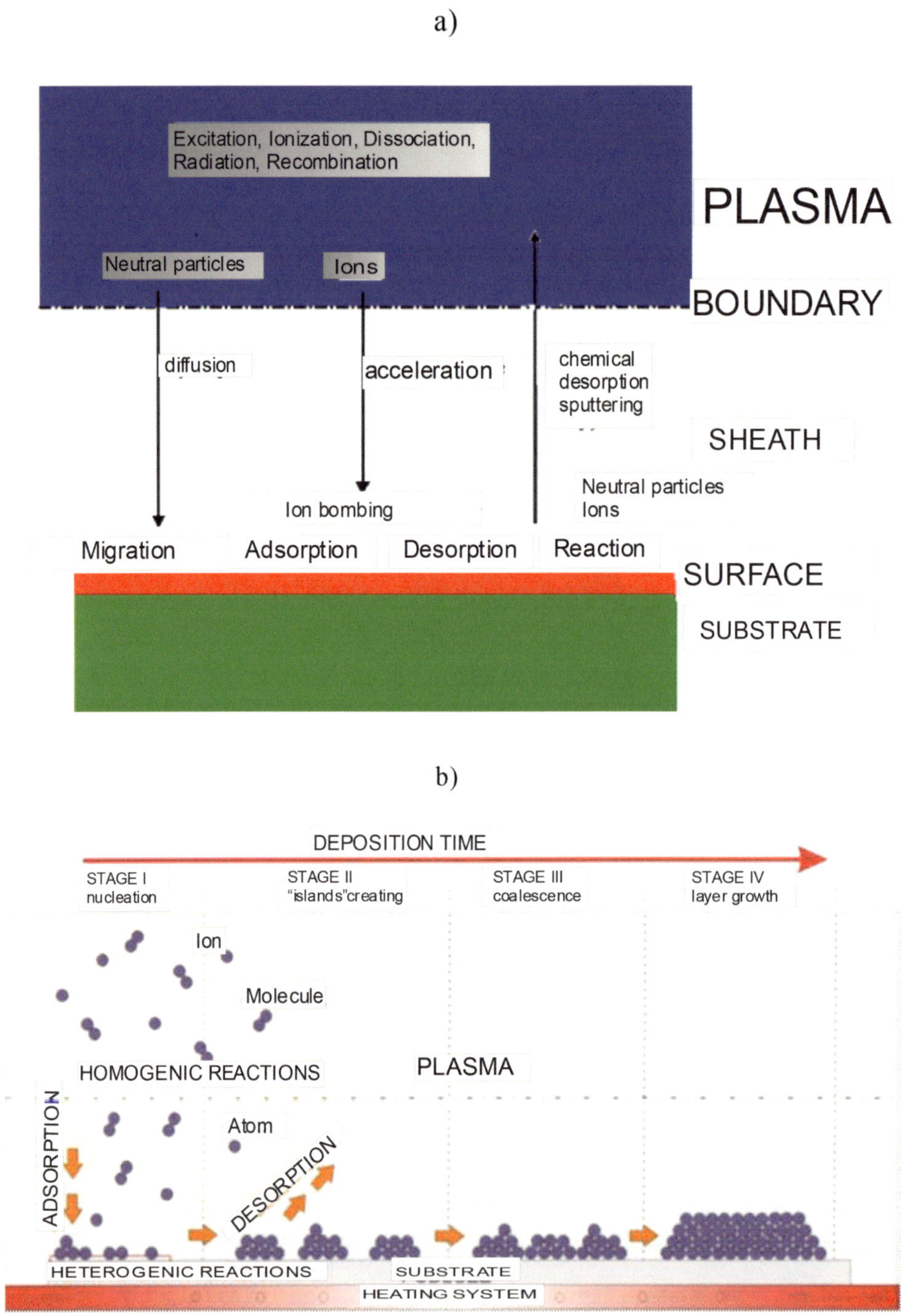

Figure 3. (a) Scheme of reactions occurring in the PACVD process. (b) Scheme of processes occurring in the plasma reactor with the stages of coating deposition. (according to Kyziol's work [18]).

3. Techniques for obtaining DLC layers

3.1. Overview of techniques for receiving thin layers

Techniques for obtaining DLC layers can be divided into two groups [19]. The criterion for this division is the type of precursor used in the process.

The first group brings together methods in which the precursor is solid carbon:

- Atomizing a carbon electrode with a single-ion beam (sputtering beam)—spraying a carbon disk with a single argon ion beam with energy from 0.5 to a few keV; the carbon atoms removed from the electrode after getting bombarded with gas ions are ionized themselves and deposited on the appropriate surface, forming an amorphous layer without the presence of hydrogen [20] (**Figure 4a**).
- Atomization of the carbon electrode with a double-ion beam (dual-ion beam sputtering)—introducing of an additional source of hydrogen or hydrocarbon ions with 1 keV energy; their task is to bombard the growing carbon layer, after which carbon ions react with hydrogen or hydrocarbon to form a hard layer with a quasi-amorphous structure (**Figure 4b**).

The second technique includes methods in which saturated or unsaturated hydrocarbon gas is used as a precursor (acetylene, benzene, butane, cyclohexane, ethane, ethylene, hexane, isopropane, methane, pentane, propane, and propylene). This group of methods consists primarily of plasma-assisted chemical vapor deposition (PACVD) technologies. These include methods such as:

- MW CVD (microwave chemical vapor deposition)
- RF CVD (radio-frequency chemical vapor deposition)
- DC CVD (direct-current chemical vapor deposition)
- Arc discharge method

Currently, the methods included in the second group receive the most interest and are subject to continuous and intensive modernization.

There is also a method that cannot be assigned to the abovementioned groups in any way. It is a method of laser graphite ablation involving the evaporation of a graphite target using a laser.

Due to the circumstances (CVD methods described above), I will discuss this second group in more detail, that is, obtaining DLC layers using PA CVD methods.

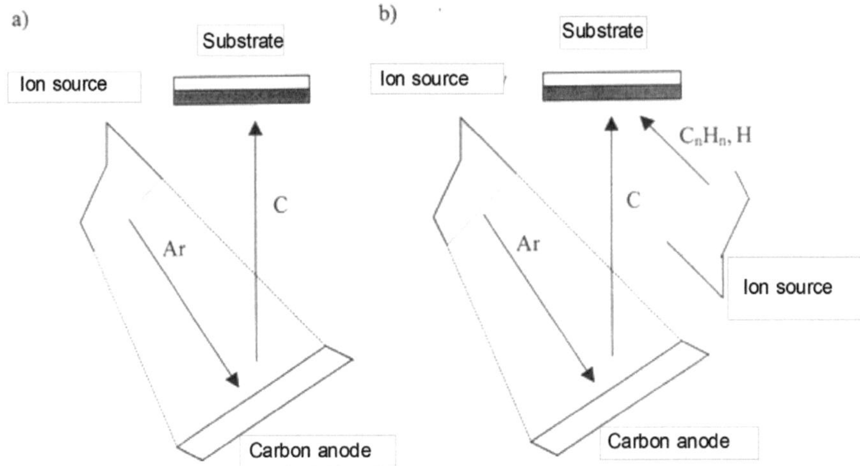

Figure 4. Scheme of systems for deposition by ion spraying: (a) single ion beam, (b) dual ion beam [21].

3.2. PA CVD and DLC layers

In the twenty-first century, most research pertains to the technology of DLC coatings, including the modification of these other atoms (in medical applications N, Si, Ti, O, F, and Ag) [22–30]. For their synthesis, the most common method is chemical vapor deposition.

The discussed layers, due to their number of interesting properties (**Table 1**), are used in many branches of the industry.

In "Introduction," I already mentioned pioneer studies concerning these layers. Up to now, however, carbon nitride has not been obtained in a crystalline form with stoichiometric or a very similar atomic ratio, and a series of works carried out in this area led to obtaining carbon layers subsidized with nitrogen atoms (N-DLC) [38–40], silicon (Si-DLC) [25, 30, 41–43], or more complex, in terms of chemical composition, SiCN:H structures [44, 45].

DLC coatings are largely amorphous, in which carbon is present in the sp^2 and sp^3 hybridizations. The clusters of atoms have a structure typical for graphite, somehow immersed in an amorphous matrix, which consist of carbon atoms [46]. The general classification of the discussed coatings is as follows [31, 47]:

- Hydrogenated structures
- Nonhydrogenated structures

In both cases, they can be doped with both metals and nonmetals. Many scientists have been investigating layers with various types of atoms [48–54].

The addition of Ti, W or Si, and N improves adhesion to the substrate, and according to Galvan et al. [49], they decrease the level of their own stresses in the DLC structure. In turn,

Type of material	sp^3 (%)	H (%)	Density (g/cm^3)	Hardness (GPa)
Diamond	100	0	3.5	100
Graphite	0	0	2.3	
a-C:H (soft)	60	40–50	1.2–1.6	<10
a-C:H (hard)	40	30–40	1.6–2.2	10–20
ta-C:H	70	30	2.4	50
N-DLC	14–21	About 36	1.9–2.4	22–28

Table 1. Selected properties of carbonaceous materials, including DLC and N-DLC layers (based on[18, 31–37]).

Dwivedi et al. [50] confirmed that nitrogen, as an additive (N-DLC), causes changes in the morphology of coatings both on a nanometric and micrometric scale, which favors the binding of sp^2 hybridization clusters. The improvement of anticorrosive properties by doping with silicon atoms was noticed by Kim et al. [51], even with the Si contribution not exceeding 2% at. In addition to the abovementioned influence, Si-DLC coatings also exhibit a more favorable hardness—hindering the formation of sp^2 moieties for the contribution of amorphous structure to sp^3 hybridization (e.g., a decrease in hardness from 23.1 to 13.7 GPa, with an increase in Si content from zero to 19.73% at [48]) [54] in comparison to, among others, Si/N-DLC coatings, high growth rate in the production process [52], good oxidation resistance, and low value of the coefficient of friction [53]. Thanks to these properties, these coatings have a wide range of applications in the automotive, aerospace, and shipbuilding industries.

Therefore, it can be argued that for the design of technology based on chemical reactions, in short CVD, the key step is the optimization of the structure of the obtained layers and the related functional properties of the materials. The goal of this approach to the discussed issue is to ensure the best possible adjustment of the surface treatment parameters in terms of the use of a given structural element.

3.3. The influence of PA CVD process parameters on the structure of N-DLC coatings

Proper selection of the technological process parameters for coating production is one of the key factors that improves their synergism on various types of substrates. In the case of chemical plasma vapor deposition (PA CVD (plasma-assisted chemical vapor deposition)) processes, this is particularly important. The keys to the construction and physicochemical properties of a given surface are:

- Duration of the process
- Temperature in the reactor and at the surface itself
- Pressure in the reaction chamber

- Power of the plasma generator
- Flow rate and type of gas precursor
- Condition and type of surface

In this respect, the technology is extremely difficult to develop. The processes occurring during deposition largely take place away from thermodynamic equilibrium. Summing up the design and development of this technology most often requires a number of experimental studies.

The research conducted by our team on the impact of gas mixture composition, process temperature, plasma generator power, and reactor pressure on the growth rate of N-DLC structures was a valuable guideline for its design. With regard to nitrogen-rich DLC layers, it can be concluded that the growth rate of these structures decreases with increasing temperature and as the proportion of N_2 in the gas mixture increases. This can be counteracted by increasing the power of the plasma generator and the pressure of the gaseous reactants in the working chamber. It is worth noting that for each type of layers, including N-DLC, limit values of individual parameters are defined, at which the aforementioned layers are thermodynamically stable [55].

4. DLC layers on different substrates

Adhering strictly to the guidelines presented in this chapter, in designing technology for modification of all kinds of materials, the parameters for individual stages of machining were selected so as to minimize the possibility of ionic etching of the resulting coatings to ensure maximum speed of the deposition process.

Figure 5 schematically represents the structure of the layers on Ti6Al4V.

The following subsections present examples of DLC layer deposition processes on several types of substrates, and the parameters of which are presented in **Table 2**.

4.1. Titanium and aluminum alloys

The diffusion processes are important for the treatment technology performed on these Ti- and Al-based alloys, which consequently leads to the formation of phases that strengthen the surface layer. In the case when their surface is nitrided, this process is usually carried out in N_2 [57, 58], N_2-H_2 [59, 60], and NH_3-Ar atmospheres [61, 62]. While the role of nitrogen is obvious, the effect of hydrogen has been explained in various ways. For example, Matsumoto [63] believes that H_2 reduces the reaction rate, while Renevier et al. [64] assume that the addition of H_2 to N_2 is not important. On the other hand, Hudis [65] demonstrates that the addition of hydrogen ensures a more effective cleaning of the surface, which promotes the nitriding process. In addition, studies conducted by Negm [66] have confirmed that the replacement of hydrogen with argon or another inert gas reduces the thickness of diffusion layers and the hardness of the treated surface.

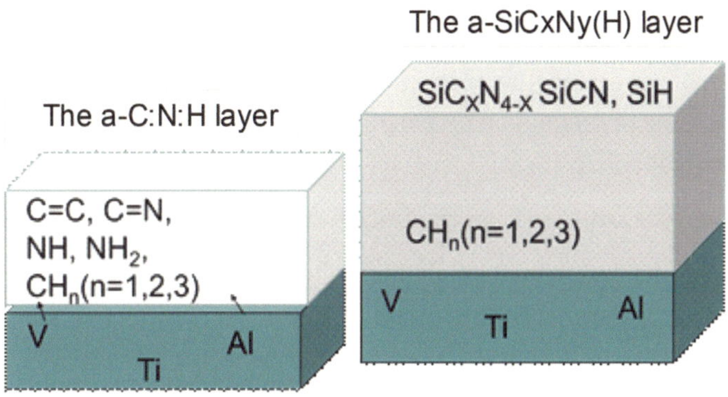

Figure 5. The scheme of DLC layer structure.

Substrate	Type of plasmochemical process	Technological parameters				
		Gas/flow (sccm)	$P_{RF/MW}$ (W)	T_s (°C)	p (Tr)	t (min)
Ti Grade2	SiCH	Ar/225; CH_4/25; SiH_4/3	P_{MW} 400	600	0.3	30
	SiNH	Ar/225; NH_3/150; SiH_4/3	P_{MW} 400	600	0.3	30
	SiCNH	Ar/225; NH_3/130; CH_4/20; SiH_4/3	P_{MW} 400	600	0.3	30
Al-Zn alloy	N^+ ion	N_2/90; H_2/30	P_{RF} 100	350	0.7	120
	Si-DLC	Ar/80; SiH_4/8; CH_4/8	P_{RF} 50	400	0.4	8
Polyetherketone	CNH	Ar/75; N_2/85; CH_4/10	P_{RF} 80	24	0.4	30
	SiCNH	Ar/75; N_2/25; CH_4/104; SiH_4/3	P_{RF} 80	24	0.4	30

$P_{RF/MW}$, radio-frequency or microwave generator power; T_s, substrate temperature; p, pressure in the reactor; t, time of process deposition.

Table 2. Details for plasma modification including carbon-based coating deposition [56].

Our team conducted a lot of researches on DLC layers deposited on substrates [56]. The results for the Ti Grade 2 (after plasma processes) obtained from the biological and corrosion tests suggest that titanium with the plasmochemically modified surface did not cause severe cytotoxicity against *MG63* cells and improves corrosion properties.

Similarly, beneficial results of the transfer of stresses between the substrate and the coating after using the surface modification process with nitrogen or silicon ions were obtained in the work on the modification of aluminum alloys [43, 67, 68]. For example, Lifang et al. [69] clearly indicate that obtaining an intermediate layer, at the Al-Cu-Mg-DLC alloy boundary, decisively influences, inter alia, the hardness of the surface exposed to working conditions. The various intermediate layers obtained by these authors, under different conditions of plasma ion implantation (PBII (plasma-based ion implantation)), lead to changes in the degree of surface hardening (in the range 17–22 GPa) and the extent of the alloy's strengthening (800–1350 nm).

However, with respect to the structure of the coatings themselves, for example, Wang et al. [70] and Ikeyama et al. [71] confirmed that the addition of Si to the DLC structure improves the adhesion of the coating to the substrate and reduces the level of internal stresses in its structure. In turn, subsidizing the DLC structure with nitrogen atoms results in a decrease in the coefficient of friction [72].

In the case of our team's research on modification of the Al-Zn alloy (after modification of nitrogen ions and deposition of SiCH layers), an improvement in hardness (up to 8 GPa), Young's modulus (up to 6 GPa), and surface smoothness (surface roughness parameter) were observed.

Figure 6 presents photos of modified metallic materials.

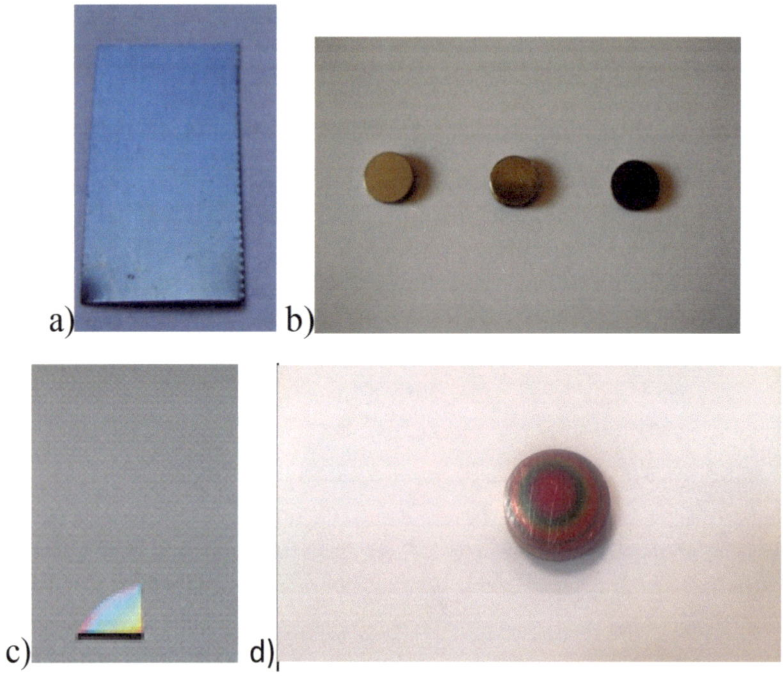

Figure 6. Images of different substrates with DLC layers. (a) Ti Grade 2, (b) Ti6Al4V, (c) Ti-Al, and (d) Al-Zn.

4.2. Polymers

During the last decade, PEEK has generated much interest thanks to its thermoplastic properties, chemical resistance, high-temperature stability, and biocompatibility [73, 74]. It has been considered as a suitable material replacing metals in biomedical implants and as a matrix of carbon fiber composites [75, 76]. However, it also has disadvantages, namely, hydrophobicity and low surface free energy, resulting in poor adhesive properties. In order to improve the properties of PEEK, we modified the samples with ion etching and deposited layers a-CN:H/a-SiCN:H by using RFCVD [56].

Figure 7 presents photo of modified PEEK.

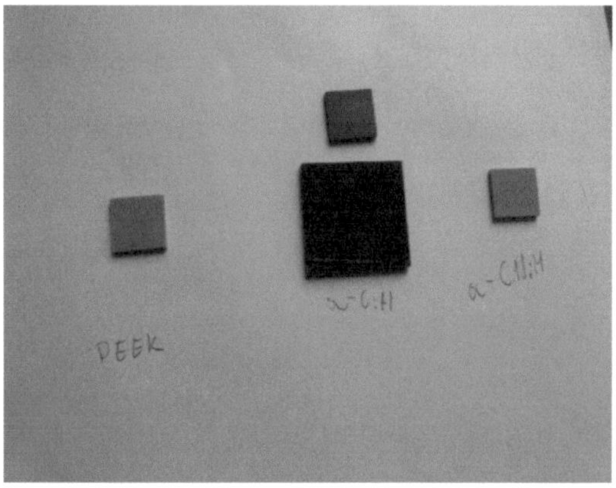

Figure 7. Image of PEEK with DLC layers.

4.3. Austenitic-ferritic steel

In this respect, especially in the case of DLC coatings obtained on metallic substrates (e.g., steel), this process is preceded by interlayer deposition. A transitional layer often containing silicon [43] significantly improves the adhesion of the specific coating to metal alloys and at the same time prevents an unfavorable phenomenon, which in this case is the precipitation of iron-carbon phases.

In the work of Stypuła, et al. [77] it was found that nitriding austenitic-ferritic steel (duplex) using radio-frequency plasma, under low-pressure and low-temperature conditions ($p = 0.7Tr$, $T = 573$ K) improves the passivation parameters and corrosion resistance of the LOOH21N7MC alloy. This increase in corrosion resistance is associated with high nitrogen content (~21%), with the oxide-nitride structure of the nitrided outer layer, as well as a high degree of surface enrichment in the silicon in the form of a nitride.

Brylewski et al. [78] used PACVD to obtain a silicon carbonitride layer with a defined desired chemical composition on Crofer 22APU steel, both without and after prior nitriding. It was determined that nitriding has a positive influence on the microhardness and elasticity modulus of the investigated $SiC_xN_y(H)$ layer. High-temperature oxidation studies performed on the samples indicate that prior nitriding in plasmochemical conditions results in improved corrosion resistance of the Crofer 22APU steel with the deposited $SiC_xN_y(H)$ layer. From these studies it can be concluded that the deposition of the silicon carbonitride layer preceded by nitriding can be successfully used to obtain materials with potential applications as interconnects in planar-type intermediate-temperature solid oxide fuel cells.

5. Application

Thanks to their properties, such modified materials have found a fairly wide application in several branches of the industry, construction, and medicine. The first group includes, among others, car industry, aviation, shipbuilding, electronics. The second one is mainly photovoltaics, various types of constructions, machines, and parts of construction machines. In contrast, the

Figure 8. Examples of applications for modified materials.

third group consists of a wide range of applications, from bone implants to medical instruments and veterinary. A separate group includes sports, especially golf, sailing, or canoeing.

Figure 8 presents photos of examples of applications for modified materials.

Author details

Marta Januś

Address all correspondence to: martaj@agh.edu.pl

AGH–University of Science and Technology, Faculty of Materials Science and Ceramics, Krakow, Poland

References

[1] Burakowski T. Rozważania o synergizmie w inżynierii powierzchni [Considerations on Synergism in Surface Engineering]. Radom: Wydawnictwo Politechniki Radomskiej; 2004

[2] Musil J, Vlček J, Růžička M. Recent progress in plasma nitriding. Vacuum. 2000;**59**:940-951

[3] Chu PK, Chen JY, Wang LP, Huang N. Materials science and engineering reports: A review. Journal. 2002;**36**:143-206

[4] Liu AY, Cohen ML. Structural properties and electronic structure of low-compressibility materials: Beta-Si_3N_4 and hypothetical beta-C_3N_4. Physical Review. 1990;**B41**:10727

[5] Matsumoto O, Kotaki T, Sikano H. Synthesis of carbon nitride in plasma arc. Journal of the Electrochemical Society. 1994;**141**:L16-L18

[6] Yu KM, Cohen ML, Haller EE. Observation of crystalline C_3N_4. Physical Review. 1994;**B49**:5034-5037

[7] Xianhui B, Zheng N, Li Y, Feng P. Templated assembly of sulfide nanoclusters into cubic-C_3N_4 type framework. Journal of the American Chemical Society. 2003;**125**(20): 6024-6025

[8] Chen LC, Luang JY. Crystalline silicon carbon nitride: A wide band gap semiconductor. Applied Physics Letters. 1998;**72**(19):2463-2465

[9] Chen LC, Wei SL. Crystalline SiCN: A hard material rivals to cubic BN. Thin Solid Films. 1999;**355-356**:112-116

[10] Błaszczyk-Lezak I, Wróbel AM, Bieliński DM. Remote hydrogen microwave plasma chemical vapor deposition of silicon carbonitride films from a (dimethylamino) dimethylsilane precursor: Compositional and structural dependencies of film properties. Diamond and Related Materials. 2006;**15**:1650-1658

[11] Fainer NI, Maximovski EA, Rumyantsev YM, Kosinova ML, Kuznetsov FA. Composition and structure of silicon carbonitride layers grown on Si(100)/(Fe, N, Co) substrates. Nuclear Instruments and Methods in Physics Research A. 2001;**470**:193-197

[12] Norm ISO 5832/3-1978 (E). Implants for surgery—Metallic materials wrought titanium 6-aluminium 4-vanadium alloy. ISO; 1978

[13] Norm ASTM F136-98. Standard specification for wrought titanium-6aluminum-4vanadium ELI. USA; 1998

[14] Somiya S, Goto T. Handbook of Advanced Ceramics. Elsevier; 2013

[15] Langmuir I. Oscillations inionied gases. Physical Review. 1928;**14**:627-637

[16] Merche D, Vandencasteele N, Reniers F. Atmospheric plasmas for thin film deposition: A critical review. Thin Solid Films. 2012;**520**:4219-4236

[17] Lebrun JP. Applications of low-temperature surface hardening of stainless steels. Thermochemical Surface Engineering of Steels. Elsevier; 2015. pp. 633-647

[18] Kyziol K. Funkcjonalne powłoki otrzymywane w warunkach plazmochemicznych [Functional coatings obtained under plasmochemical conditions]. Ceramics. 2016;**123**

[19] Spear KE, Dismukes JP. Synthetic Diamond: Emerging CVD Science and Technology. USA: John Wiley & Sons, Inc.; 1994

[20] Paiva OC, Santos JD, Reis RL. An electrochemical study on anodised Ti-6Al-4V aiming at improving its biomedical performance. In: Proceedings of the Symposium on Compatibility of Biomedical Implants. Vol. 94-15. p. 145

[21] Anderson L-P. A review of recent work on hard i-C films. Thin Solid Films. 1981;**86**:193

[22] Kluba A, Bociąga D, Dudek M. Hydrogenated amorphous carbon films deposited on 316L stainless steel. Diamond and Related Materials. 2010;**19**:533-536

[23] Tétard F, Djemia P, Angleraud B, Mubumbila N, Tessier PY. Surface and bulk characterizations of CNx thin films made by r.f. magnetron sputtering. Surface and Coatings Technology. 2002;**151-152**:184-188

[24] Zhang P, Tay BK, Yu GQ, Lau SP, Fu YQ. Surface energy of metal containing amorphous carbon films deposited by filtered cathodic vacuum arc. Diamond and Related Materials. 2004;**13**:459-464

[25] Bociąga D, Kamińska M, Sobczyk-Guzenda A, Jastrzębski K, Świątek L, Olejnik A. Surface properties and biological behaviour of Si-DLC coatings fabricated by a multi-target DC-RF magnetron sputtering method for medical applications. Diamond and Related Materials. 2016;**67**:41-50

[26] Bociąga D, Komorowski P, Batory D, Szymański W, Olejnik A, Jastrzębski K, Jakubowski W. Silver-doped nanocomposite carbon coatings (Ag-DLC) for biomedical applications—Physiochemical and biological evaluation. Applied Surface Science. 2015;**355**:388-397

[27] Bociąga D, Jakubowski W, Komorowski P, Sobczyk-Guzenda A, Jędrzejczak A, Batory D, Olejnik A. Surface characterization and biological evaluation of silverincorporated DLC coatings fabricated by hybrid RF PACVD/MS method. Materials Science and Engineering C. 2016;**63**:462-474

[28] Batory D, Jędrzejczak A, Kaczorowski W, Szymański W, Kołodziejczyk L, Cłapa M, Niedzielski P. Influence of the process parameters on the characteristics of silicon-incorporated a-C:H:SiOx coatings. Surface and Coatings Technology. 2015;**271**:112-118

[29] Batory D, Jędrzejczak A, Szymański W, Niedzielski P, Fijałkowski M, Louda P, Kotela I, Hromadka M, Musil J. Mechanical characterization of a-C:H:SiOx coatings synthesized using radio-frequency plasma-assisted chemical vapor deposition method. Thin Solid Films. 2015;**590**:299-305

[30] Szmidt J, Beck RB, Lisik Z, Mitura S, Sokołowska A. State of the silicon-DLC layer interface produced by plasma methods. Diamond and Related Materials. 1996;**5**:1204-1209

[31] Robertson J. Diamond-like amorphous carbon. Materials Science & Engineering R: Reports. 2002;**37**:129-281

[32] Field JE, editor. The Properties of Natural and Synthetic Diamond. London: Academic Press; 1992

[33] Cahn RW. Physics of graphite. Journal of Nuclear Materials. 1983;**114**:116-116

[34] Koidl P, Wild C, Dischler B, Wagner J, Ramsteiner M. Plasma deposition, properties and structure of amorphous hydrogenated carbon films. Materials Science Forum. 1990;**52-53**:41-70

[35] Weiler M, Sattel S, Jung K, Ehrhardt H, Veerasamy VS, Robertson J. Highly tetrahedral, diamond-like amorphous hydrogenated carbon prepared from a plasma beam source. Applied Physics Letters. 1994;**64**:2797-2799

[36] Donnet C. Recent progress on the tribology of doped diamond-like and carbon alloy coatings: A review. Surface and Coatings Technology. 1998;**100-101**:180-186

[37] Murata Y, Choo CK, Ono H, Nagai Y, Tanaka K. Characterization of N-doped DLC thin films prepared by hydrocarbons pyrolysis method. Materials Today: Proceedings. 2016;**3**:197-202

[38] Motta EF, Pereyra I. Amorphous hydrogenated carbon-nitride films prepared by RF-PECVD in methane-nitrogen atmospheres. Journal of Non-Crystalline Solids. 2004;**338-340**:525-529

[39] Neuville S, Matthews A. A perspective on the optimisation of hard carbon and related coatings for engineering applications. Thin Solid Films. 2007;**515**:6619-6653

[40] Yang P, Huang N, Leng YX, Chen JY, Wang J, Sun H, Wan GJ, Zhao AS. Wettability and bloodcompatibility of a-C:N:H films deposited by PIII-D. Surface and Coatings Technology. 2010;**204**:3039-3042

[41] Soum-Glaude A, Thomas L, Dollet A, Ségur P, Bordage MC. Argon/tetramethysilane PECVD: From process diagnostic and modeling to a-Si:C:H hard coating composition. Diamond and Related Materials. 2007;**16**:1259-1263

[42] Iseki T, Mori H, Hasegawa H, Tachikawa H, Nakanishi K. Structural analysis of Si-containing diamond-like carbon. Diamond and Related Materials. 2006;**15**:1004-1010

[43] Cemin F, Boeira CD, Figueroa CA. On the understanding of the silicon-containing adhesion interlayer in DLC deposited on steel. Tribology International. 2016;**94**:464-469

[44] Peter S, Bernütz S, Berg S, Richter F. FTIR analysis of a-SiCN:H films deposited by PECVD. Vacuum. 2013;**98**:81-87

[45] Kafrouni W, Rouessac V, Julbe A, Durand J. Synthesis of PECVD a-SiCxNy:H membranes as molecular sieves for small gas separation. Journal of Membrane Science. 2009;**329**:130-137

[46] Spencer EG, Schmidt PH, Joy DC, Sansalone FJ. Ion beam deposited polycrystalline diamond-like films. Applied Physics Letters. 1976;**29**.118-120

[47] Lifshitz Y. Diamond-like carbon—Present status. Diamond and Related Materials. 1999;**8**:1659-1676

[48] Bogdanowicz R. Advancements in diamond-like carbon coatings. Handbook of Nanoceramic and Nanocomposite Coatings and Materials. Elsevier; 2015. pp. 183-200

[49] Galvan D, Pei YT, De Hosson JTM. Influence of deposition parameters on the structure and mechanical properties of nanocomposite coatings. Surface and Coatings Technology. 2006;**201**:590-598

[50] Dwivedi N, Kumar S, Malik HK, Rauthan CMS, Panwar OS. Influence of bonding environment on nano-mechanical properties of nitrogen containing hydrogenated amorphous carbon thin films. Materials Chemistry and Physics. 2011;**130**:775-785

[51] Kim H-G, Ahn S-H, Kim J-G, Park SJ, Lee K-R. Effect of Si-incorporation on wear–corrosion properties of diamond-like carbon films. Thin Solid Films. 2005;**482**:299-304

[52] Jiang L, Chen X, Wang X, Xu L, Stubhan F, Merkel K-H. a-SiCx:H films deposited by plasma-enhanced chemical vapor deposition at low temperature used for moisture and corrosion resistant applications. Thin Solid Films. 1999;**352**:97-101

[53] Damasceno JC, Camargo SS, Freire FL, Carius R. Deposition of Si-DLC films with high hardness, low stress and high deposition rates. Surface and Coatings Technology. 2000;**133-134**:247-252

[54] Tang J, Wang Y. Characterization of DLC-Si films prepared by RF-PECVD. In: Optoelectronics and Microelectronics (ICOM). 2012. pp. 431-433

[55] Kyzioł K, Jonas S, Tkacz-Śmiech K, Marszałek K. A role of parameters in RF PA CVD technology of a-C:N:H layers. Vacuum. 2008;**82**:998-1002

[56] Januś M, Kluska S, Kyziol K, Jonas S. Plasma assisted chemical vapour deposition–Technical design of functional coatings. Polish Academy of Sciences, Committee of Metallurgy, Institute of Metallurgy and Materials Science, Archives of Metallurgy and Materials. 2015;**60**(2):909-914

[57] Gredelj S, Gerson AR, Kumar S, McIntyre NS. Plasma nitriding and in situ characterisation of aluminium. Applied Surface Science. 2002;**199**:234-247

[58] Quast M, Mayr P, Stock HR. Plasma monitoring of plasma-assisted nitriding of aluminium alloys. Surface and Coatings Technology. 1999;**120-121**:244-249

[59] Mubarak Ali M, Ganesh Sundara Raman S, Pathak SD, Gnanamoorthy R. Influence of plasma nitriding on fretting wear behaviour of Ti–6Al–4V. Tribology International. 2010;**43**:152-160

[60] Vissutipitukul P, Aizawa T. Wear of plasma-nitrided aluminum alloys. Wear. 2005;**259**: 482-489

[61] Wang PW, Sui S, Wang W, Durrer W. Aluminum nitride and alumina composite film fabricated by DC plasma processes. Thin Solid Films. 1997;**295**:142-146

[62] Avni R, Spalvins T. Nitriding mechanisms in Ar-N_2, Ar-N_2-H2 and Ar-NH_3 mixtures in D.C. Glow discharges at low pressures (less than 10 Torr). Materials Science and Engineering. 1987;**95**:237-246

[63] Matsumoto O. Nitriding reaction of metals under plasma conditions. In: Proceedings of the International Ion Engineering Congress ISIAT'83, Kyoto; 1983. pp. 1379-1390

[64] Renevier N, Czerwiec T, Collignon P, Michel H. Diagnostic of arc discharges for plasma nitriding by optical emission spectroscopy. Surface and Coatings Technology. 1998;**98**:1400-1405

[65] Hudis M. Study of ion nitriding. Journal of Applied Physics. 1973;**44**:1489-1496

[66] Negm NZ. A study on rf plasma nitriding at a constant power in different H_2–N_2 mixtures at different temperatures. Materials Science and Engineering B. 2006;**129**:207-210

[67] Liao JX, Xia LF, Sun MR, Liu WM, Xu T, Xue QJ. The tribological properties of a gradient layer prepared by plasma-based ion implantation on 2024 aluminum alloy. Wear. 2004;**256**:840-845

[68] Liao JX, Xia LF, Sun MR, Liu WM, Xu T, Yang CR, Chen HW, Fu CL, Leng WJ. Structural characteristics of 2024 aluminum alloy plasma-based ion implanted with nitrogen then titanium. Applied Surface Science. 2005;**240**:71-76

[69] Lifang X, Zhaohui Y, Jiaxuan L. Effects of intermediate layers on the tribological behavior of DLC coated 2024 aluminum alloy. Wear. 2004;**257**:599-605

[70] Wang J, Pu J, Zhang G, Wang L. Tailoring the structure and property of silicon-doped diamond-like carbon films by controlling the silicon content. Surface and Coatings Technology. 2013;**235**:326-332

[71] Ikeyama M, Nakao S, Miyagawa Y, Miyagawa S. Effects of Si content in DLC films on their friction and wear properties. Surface and Coatings Technology. 2005;**191**:38-42

[72] Yan X, Xu T, Chen G, Yang S, Liu H. Study of structure, tribological properties and growth mechanism of DLC and nitrogen-doped DLC films deposited by electrochemical technique. Applied Surface Science. 2004;**236**:328-335

[73] Oehr C. Plasma surface modification of polymers for biomedical use, Nuclear instruments and methods. In: Physics Research Section B: Beam Interactions with Materials and Atoms. Vol. 208. 2003. pp. 40-47

[74] Kurtz SM, Devine JN. PEEK biomaterials in trauma, orthopedic, and spinal implants. Biomaterials. 2007;**28**:4845-4869

[75] Briem D, Strametz S, Schroeder K, Meenen NM, Lehmann W, Linhart W, Ohl A, Rueger JM. Journal of Materials Science: Materials in Medicine. 2005;**16**:671-677

[76] Kurtz SM. Applications of polyaryletheretherketone in spinal implants: Fusion and mations preservation. PEEK Biomaterials Handbook. William Andrew Publishing; 2011. pp. 201-220. Chapter 13

[77] Stypuła B, Stepuch-Galik A, Kowalski K, Januś M, Bernasik A. Korozja wysokostopowego staliwa azotowanego plazmowo [Corrosion of plasma nitrided cast stainless steel]. Ochrona przed Korozją. 2005;**48**(11s/A):66-71. ISSN: 0473-7733

[78] Januś M, Kluska S, Bobruk M, Brylewski T, Jastrzębski W, Zimowski S, Grzesik Z. Effect of plasma-assisted CVD surface modification of Crofer 22APU steel on its selected physicochemical properties. Ochrona przed Korozją. 2015;**58**(11):418-422. ISSN: 0473-7733

Chapter 4

Controlled Post-treatment of Thick CVD-Diamond Coatings by High-Density Plasma Oxidation

Tatsuhiko Aizawa

Additional information is available at the end of the chapter

http://dx.doi.org/10.5772/intechopen.79143

Abstract

CVD-diamond coatings were posttreated by plasma oxidation to recycle an original WC (Co) mother tool substrate for recoating and reuse. A developed RF/DC plasma oxidation system was stated together with a hollow cathode device to intensify the oxygen ion and electron densities and with a quantitative plasma diagnosis equipment. Plasma oxidation ashing conditions were optimized by this quantitative diagnosis toward the perfect ashing of diamond films with less residuals and less tool edge damage. Geometric effect of tool teeth structure on this ashing process was discussed by in situ monitoring of plasmas. Engineering solution of this ashing process was proposed for industrial applications.

Keywords: plasma oxidation, CVD-diamond coating, ashing, hollow cathode device, quantitative plasma diagnosis

1. Introduction

CVD-diamond coatings as well as poly-crystalline diamonds have been widely utilized for the protection of dies and tools from wear and damage in cutting, drilling, and piercing operations in practice [1]. They are also expected to be working as a semi-conductive and thermal-conductive substrate for micro-electro-mechanical system (MEMS) and Bio-MEMS [2]. In parallel with high qualification in CVD-diamond coating technologies, how to make accurate shaping, micro−/nano-texturing, polishing and finishing of diamond films also becomes an engineering issue [3].

In the literature of patents to search for qualification in ashing the used diamond coatings, many studies have been developed to efficiently remove the diamond films from WC (Co) (Tungsten Carbide with Cobalt binder) and silicon substrates. Among them, three ashing processes were noticed; the laser machining [4], the ion-beam and plasma discharging [5], and the reactive etching method [6]. The former two methods are difficult to be applied to tools with complex shaped teeth. Residuals of diamond were left on the treated tool surface. WC (Co) substrate suffered from severe thermal and chemical transients. Plasma oxidation method [7] has grown up as one of the most promising approaches to ashing. The CVD-diamond films are physically or chemically removed by ion bombardment and/or plasma-driven chemical reactions. These plasma ashing processes encounter some severe practical difficulties: (1) imperfect ashing of diamond films, (2) long leading time, (3) residual diamond films, especially behind the cutting teeth, (4) overoxidation of WC (Co) substrate, and (5) damage at cutting tooth edges. The used carbon-base coatings were removed from substrates by nontraditional plasma oxidation method for industrial application [8–11]. In this chapter, RF/DC plasma oxidation system with hollow cathode devices is introduced as an effective ashing method. Plasma oxidation processing for posttreatment of diamond films is stated with comments on its fundamental features and application to industries. Quantitative plasma diagnosis is utilized to characterize the oxygen plasmas and to describe the ashing process. Two types of CVD-diamond-coated cutting tools are employed as a specimen to demonstrate the effectiveness of present plasma oxidation ashing.

2. Plasma oxidation processing

The glow discharging process is utilized to ignite the oxygen plasmas by the application of radio-frequency (RF) power and direct current (DC) bias. This oxygen plasma density is effectively intensified by using the hollow cathode devices. Emissive light optical spectroscopy (EOS) as well as the Langmuir probe are utilized to quantitatively characterize the oxygen plasma state and to describe the reaction process during oxidation. CVD-diamond-coated tools are prepared for ashing treatments.

2.1. RF/DC plasma oxidation system

Different from the conventional DC- or RF-plasma generators, where plasmas are ignited and generated in the frequency of 13.56 MHz or its multiples, the present RF/DC oxygen plasmas are ignited at 2 MHz without use of matching box, as shown in **Figure 1**.

This frequency is automatically adjusted to keep matching between input and out powers around 2 MHz [7, 12–13]. This automatic and prompt response in power matching stabilizes the temporally varying plasma state. As reported in [7], a longer response time in the order of 1–10 s is needed for the mechanical matching box to adjust the RF-power. The present oxygen plasma state for ashing process is controlled within 1 ms in response; there is no time delay in power control. The vacuum chamber is electrically neutral without a self-biased current. Both the RF-voltage and DC-bias are controlled independently from each other to be from 0 to 250 V, and, from 0 to 600 V, respectively. This RF-plasma is ignited and controlled by using a dipole electrode, while the DC bias is directly applied to the specimens. In the following ashing experiments, the specimens were located on the cathode table before evacuation down to the base pressure. Then, a carrier gas was introduced into the chamber to attain the specified pressure. Both oxygen and argon gases were available in this system besides the nitrogen gas for bent.

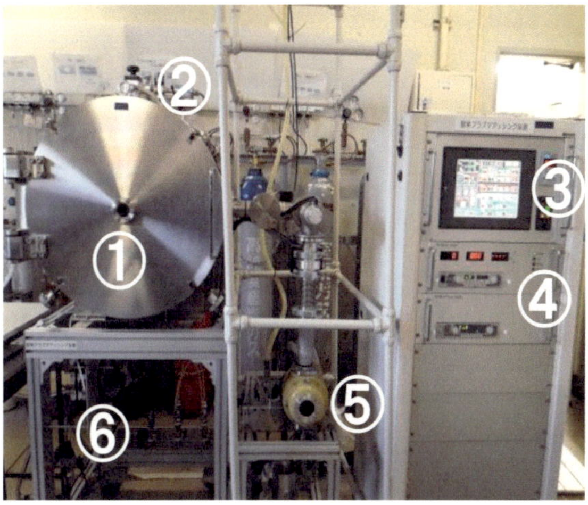

Figure 1. High density RF-DC oxygen plasma etching system. ① vacuum chamber, ② RF-generator, ③ control panel, ④ RF- and DC-power supplies, ⑤ evacuation units, and ⑥ carrier gas supply.

2.2. Plasma oxidation diagnosis

EOS (Hamamatsu Photonics, Co., Ltd., Japan) is first utilized for diagnosis on the oxygen plasmas, as shown in **Figure 2**. The emissive light from plasmas is detected through a silica window of chamber to send the signals to the analyzer via the optical fibers. This spectroscopic analysis is also useful to make on-line monitoring on plasma state in ashing. Each activated species in the plasma is identified by its corresponding peak at the specified wave length in the measured spectrum. **Figure 2** also illustrates the Langmuir probe, equipped to the present ashing system. Both the ion/electron density and electron temperature are measured for optimization of ashing process. With this information of density and temperature

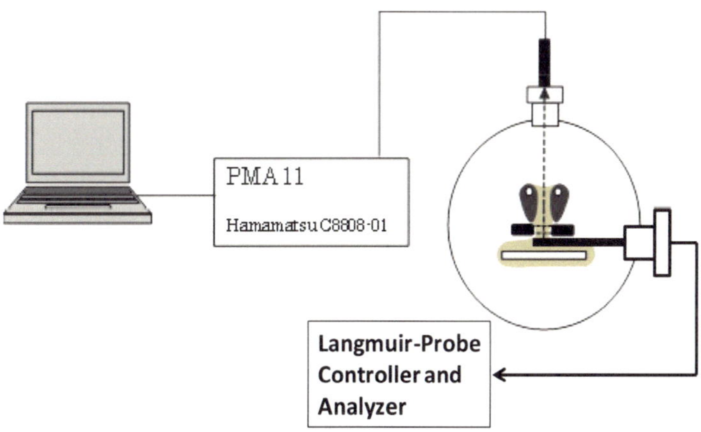

Figure 2. Quantitative plasma diagnosis to measure the electron density and temperature as well as the spectrum of activated species in plasmas.

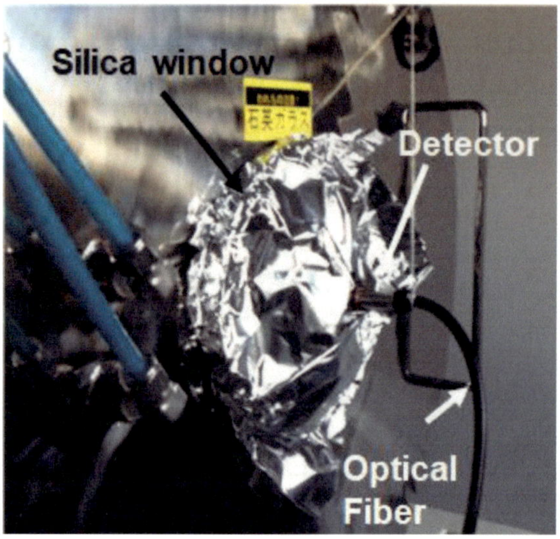

Figure 3. An experimental set-up for plasma diagnosis by EOS.

both for electrons, ions and radicals, the plasma state in processing are precisely defined and described to investigate the effect of plasma processing parameters on the ashing behavior.

Let us first explain the experimental set-up for EOS and measure the typical spectra in the function of oxygen gas pressure [14]. **Figure 3** shows an experimental set-up to make EOS. The whole window was covered by a pure aluminum sheet to be free from noises. The measuring point can be varied by focusing; the plasma state at the focused point in the chamber is in situ measured during oxidation process.

When RF voltage was 250 V and DC-bias, −500 V, the emissive light optical spectra were measured in the function of gas pressure. As shown in **Figure 4**, the oxygen molecular ion O_2^+ as well as the oxygen ion (OII or O^+) has the highest peak at 774 and 470 nm, respectively.

The intensity of oxygen ions (OI or O*; OII or O^+; OIII or O^{2+}) and O_2^+ reduced with increasing the gas pressure in **Figure 4**. This is because the reaction cross-section significantly reduces for electron detachment from oxygen molecules.

The Langmuir probe is employed to measure the electron and ion densities as well as the electron temperature. The experimental set-up for I-V curve measurement in the RF/DC plasmas is depicted in **Figure 5**. The accuracy in measurement strongly depends on the tip geometry and resistivity; e.g., the tip radius is 350 μm, the tip holder, 3 mm, the probe tip length, 1 cm, and, the tip resistance, 36 Ohm, as shown in **Figure 5(b)**.

The measured I-V curve is nearly constant but abruptly changes at the vicinity of plasma potential. This I-V curve is deconvoluted to deduce each contribution of ions and electrons to measured current density. Each I-V curve for ion and electron is differentiated to calculate its first and second derivatives across the plasma potential; the electron and ion density as well as the electron temperature are deduced from this direct calculation. **Figure 6** depicts the variation of measured ion density with the oxygen gas pressure when the DC bias was −600 V. Reduction of ion density

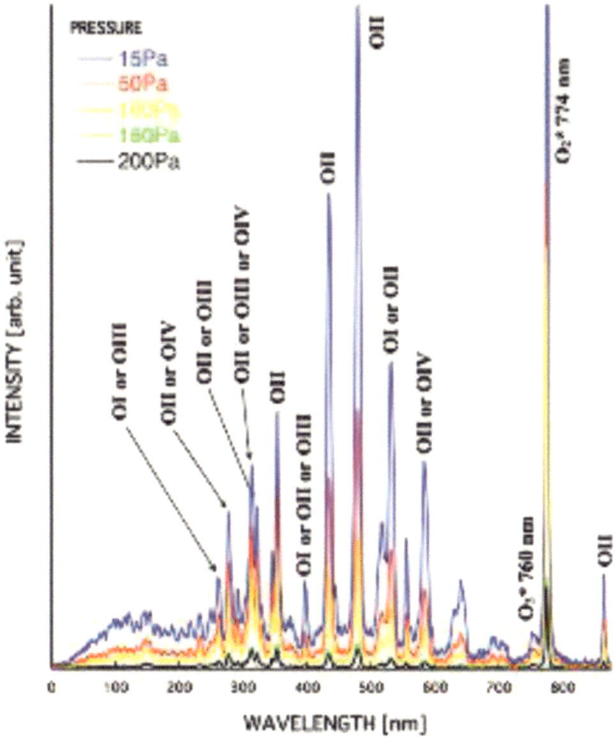

Figure 4. Measured emissive light optical spectra in the function of gas pressure for a standard RF/DC plasma under [250 V; −500 V].

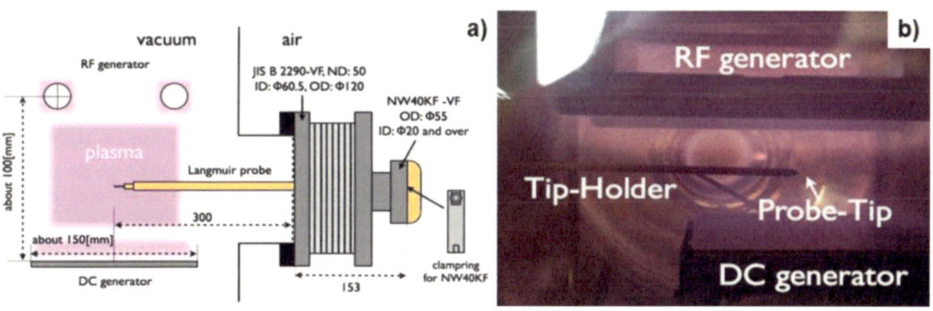

Figure 5. Experimental set-up for measurement of ion and electron densities as well as electron temperature at the specified position in the RF/DC plasmas by using the Langmuir probe.

was only measured in the range of 150–175 Pa; ion density monotonically increases with the pressure when DC bias was more than −500 V. Ionization is activated with increasing the pressure because of collision of electrons with oxygen molecules. Finally, the present RF/DC plasma characteristics is compared with the ICP (Induction-Coupled Plasma) where a cylindrical tube with diameter 40 mm and a length of 600 mm was used to ignite the plasmas. At the pressure of 75 Pa, the electron density (Ne) reaches to 4.8×10^{16} m^{-3} with the electron temperature (Te) of 3.9 eV in

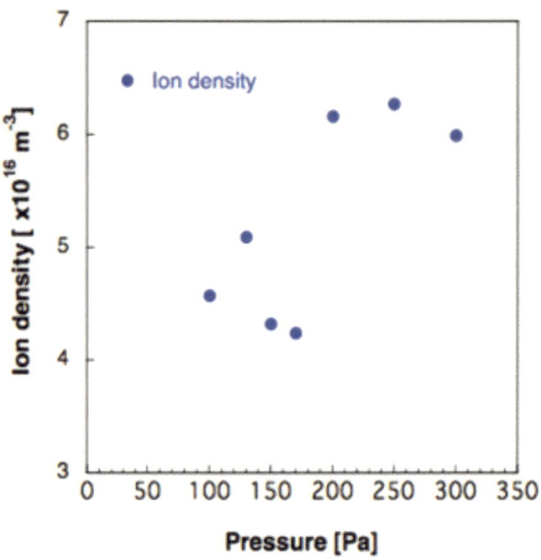

Figure 6. Variation of the measured ion density with the gas pressure by the Langmuir probe.

the present system. When using ICP, Ne = 1.0 × 10^{16} m^{-3} and Te = 5.0 eV in [15]. Higher electron density by five times can be attained even at the center of large chamber in **Figures 1** and **5(b)**.

2.3. High-density plasma oxidation

Various tools and devices are available to intensity ion and electron densities in the plasmas. A magnetism works as a lens to concentrate the ions in the magnetic field [16, 17]; ion density increases with increasing the magnetic field strength. Electric insulation is also effective to separate the plasmas into the specified spaces [18]. In this study on the plasma oxidation, a hollow cathode device [19] is selected and redesigned to confine the oxygen plasmas in the hollow and to intensity their ion and electron densities. The quantitative plasma diagnosis in 2.2 is also utilized to describe how much the oxygen ion and electron densities are enhanced by the hollow cathode device.

Variation of the electron and oxygen ion densities with pressure is compared in **Figure 7** without and with use of the hollow cathode. At the pressure of 75 Pa, N_i = 7.8 × 10^{16} m^{-3} in **Figure 7(a)**; this N_i is enhanced to be 5 × 10^{17} m^{-3} when using the hollow cathode device in **Figure 7(b)**. In the following, this hollow cathode device is redesigned to make effective ashing of diamond films coated on the cutting and end-milling tools.

2.4. Observation and measurement

High resolution optical microscopy as well as scanning electron microscopy (SEM) are utilized to observe the ashed surfaces of diamond-coated specimens. Raman spectroscopy is also used to detect a characteristic peak to the diamond.

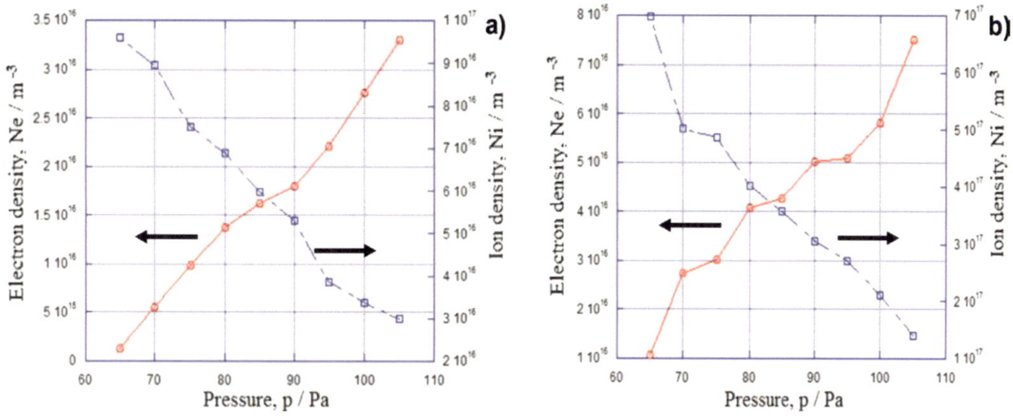

Figure 7. Comparison of the ion and electron density variation with pressure with and without use of the hollow cathode device. (a) RF-DC plasmas in **Figure 1** without use of the hollow cathode, and, (b) with use of the hollow cathode.

3. Plasma oxidation ashing of used diamond films

Plasma oxidation ashing process is employed to make perfect removal of used diamond films, coated onto the WC (Co) cutting tools, with minimum damage to tool teeth and to recycle the mother WC (Co) substrate for recoating and reuse. As an engineering issue, no residuals of diamond films are left on the WC (Co) substrate, less damage is induced during ashing process, and, this ashing must be done in short duration.

3.1. Background

CVD-diamond-coated WC (Co) cutting, drilling, and end-milling tools have been widely utilized for dry machining of the carbon fiber–reinforced plastic (CFRP) or Carbon fiber–reinforced thermoplastic (CFRTP) components and parts in the airplanes and automobiles [20, 21]. These materials are used for main cabin and wings; a hundred thousand holes must be machined or drilled into these CFRP/CFRTP structural members for each airplane in dry by using the diamond-coated tools. Since the number of airplanes is expected to be doubled in next 8 years, the above tooling cost percentage significantly increases in the total production cost. Event at present, the carbon fibers in CFRP/CFRTP have high strength and stiffness enough to make fatal damage and defect to the diamond coating of machining tools. Since an equivalent number of holes to be machined are estimated to be 100,000 per an airplane, the production cost, especially the tooling cost, increases with demanded numbers of planes or cars. Those new materials are too hard to machine or drill without damages to CVD-diamond-coated WC (Co) tools; the used tool is exchanged with new one in every machining by 10 m, or in every drilling by 10 holes [22]. How to recycle the WC (Co) substrate is essential to reduce the cost.

These used or damaged diamond coatings must be once removed to recycle the WC (Co) tool substrate with less damages to its tooth geometry before recoating and to reuse this in cutting operations. RF/DC plasma oxidation process was demonstrated to be useful for perfect removal of used tetragonal amorphous carbon coating or t-a:C coatings and for recycle of WC (Co) cutting substrate [23–25]. The recoated WC (Co) tools were proven to have the same capability in cutting performance as the virgin tools [24, 25]. This plasma oxidation process is further advanced to remove the used diamond coatings for recycle and reuse and to reduce the production cost in machining of CFRP/CFRTP members [26–28].

3.2. Preparation of diamond-coated tools

The diamond-coated WC (Co) end-milling tools were prepared for ashing experiments. As shown in **Figure 8**, two types of tools were employed to investigate the effect of tooth geometry on the plasma ashing process. The tool with the length of 70 mm, the diameter of 10 mm, and the edge skew angle of 30° was used as a standard specimen to search for an optimum ashing condition. A small-sized tool with the length of 70 mm, the diameter of 6 mm, and the edge skew angle of 10° was also used to investigate whether the optimal ashing condition is affected by down-sizing. The diamond film thickness was common in both tools by 15 μm.

3.3. Experimental set-up

An end-milling tool has skewed tooth geometry; diamond films are coated even on the curved surfaces backward the teeth. A special hollow cathode is designed to preserve the confined plasma in the hollow against the rotating tools. **Figure 9** depicts the developed hollow cathode device for ashing.

The generated RF-plasma is confined in this hollow so that higher ion and electron densities are attained for ashing the diamond films. As shown in **Figure 9(a)**, a diamond-coated tool is placed in the center of hollow tube; the same DC-bias is applied to both the hollow and the tool. Although the rotating speed has influence on the plasma configuration, it is selected to be 6 rpm in the following experiments.

A standard experimental set up is summarized as follows. After placing the specimen in **Figure 9(b)**, the chamber is evacuated down to the base pressure, less than 5×10^{-3} Pa. The pure oxygen gas with its purity of 99.99% is introduced as a carrier gas. RF- and DC- bias voltages

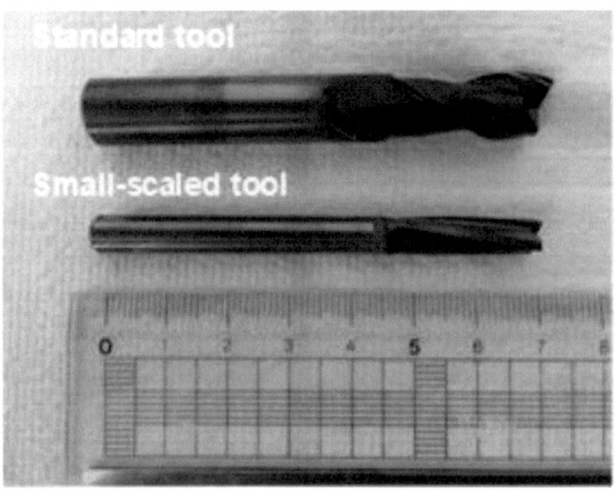

Figure 8. Two diamond-coated end-milling tools as a specimen for plasma ashing process.

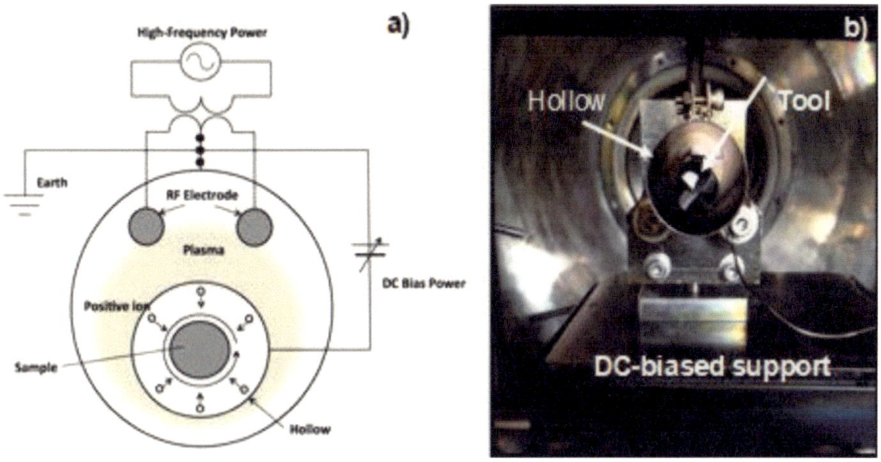

Figure 9. Special hollow cathode device for ashing the diamond films on the WC (Co) end-milling tool substrate. (a) Illustration of hollow cathode device, and, (b) experimental set-up for ashing.

are variable from 100 to 250 V and from −400 to −800 V, respectively. The plasma pressure (p) is marginable to be determined; in the following, 25 < p < 100 Pa. The plasma oxidation conditions are optimized by the quantitative plasma diagnosis to be 100 V for RF-voltage, −500 V for DC-bias, 45 Pa for pressure and 3.6 ks for duration time.

3.4. Experimental results

A standard specimen is first employed to investigate the effect of plasma processing parameters on the ashing behavior. Under the optimal conditions, the small-scaled tool specimen is also used to describe the geometric effect of tool teeth on the ashing process. Difference in ashing process conditions is explained by the in situ plasma diagnosis.

3.4.1. Standard end-milling tools

Both RF-voltage and pressure are varied to describe the plasma ashing process. In the first experiment, RF-voltage was set to be 250 V, DC-bias, −650 V and oxygen gas pressure, 30 Pa. **Figure 10** shows the ashed specimen for 3.6 ks.

Since cobalt in WC (Co) is mainly oxidized, the tool surface is colored in green in **Figure 10(a)**. No residual diamond debris are left on the tool surface in **Figure 10(b)**. Due to this excess in oxidation, the shrinkage of tool edge reaches to 7.7 μm in **Figure 10(c)**. In order to reduce the oxidation rate, the above processing parameters are tuned to be (250 V; −500 V; 45 Pa; 3.6 ks). **Figure 11** depicts the surface condition and SEM image of ashed specimen.

The surface is colored in reddish yellow in **Figure 11(a)**; this implies that a part of tungsten in WC (Co) is mainly oxidized even under this condition. Almost all the diamond films are removed on the tool surface in **Figure 11(b)**. This over-oxidation results in a significant shrinkage of edge by 9.7 μm in **Figure 11(c)**. To be discussed in later, the activated oxygen atom density must be directly controlled by lowering the RF-voltage.

Figure 12(a) depicts the ashed tool surface by lowering the RF-voltage down to 100 V. No oxide layers of cobalt and tungsten are seen on the WC (Co) substrate surface. The original diamond

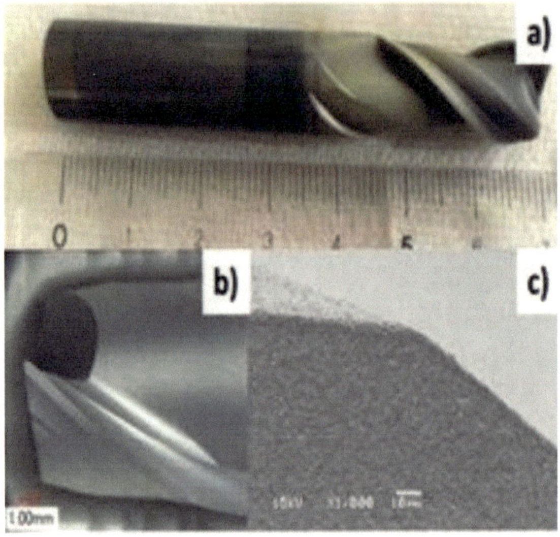

Figure 10. Ashing of CVD-diamond coating on the WC (Co) standard specimen under (250 V; −650 V; 30 Pa; 3.6 ks). (a) Surface of ashed specimen before cleansing, (b) Ashed tool edge, (c) SEM image of tool edge.

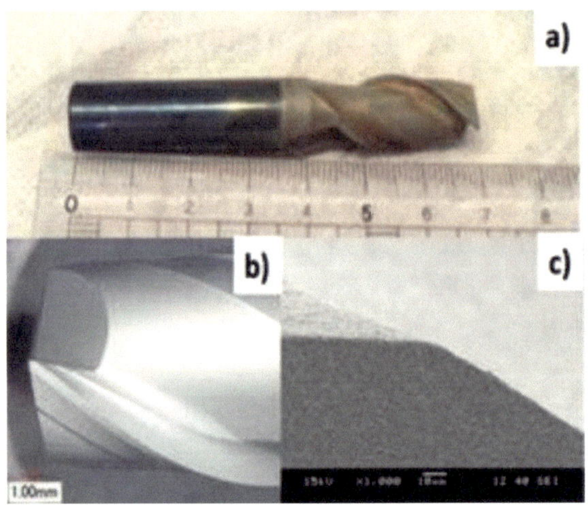

Figure 11. Ashing of CVD-diamond coating on the WC (Co) standard specimen under (250 V; –-500 V; 45 Pa; 3.6 ks). (a) Surface of ashed specimen before cleansing, (b) Ashed tool edge, (c) SEM image of tool edge.

coatings are completely removed only after ashing for 3.6 ks, as shown in **Figure 12(b)**. The shrinkage of tool edge is limited by 1.1 μm in **Figure 12(c)**. Fine SEM observation and the Raman spectroscopy are utilized to precisely describe this ashing process. **Figure 13** compares the top view in SEM image for the WC (Co) tool surface before and after ashing under this optimal condition of (100 V; –-500 V; 45 Pa; 3.6 ks).

The original tetrahedral diamond crystals with significant roughness in **Figure 13(a)** are completely removed to leave a flat WC (Co) surface with micro-pores among WC grains. These

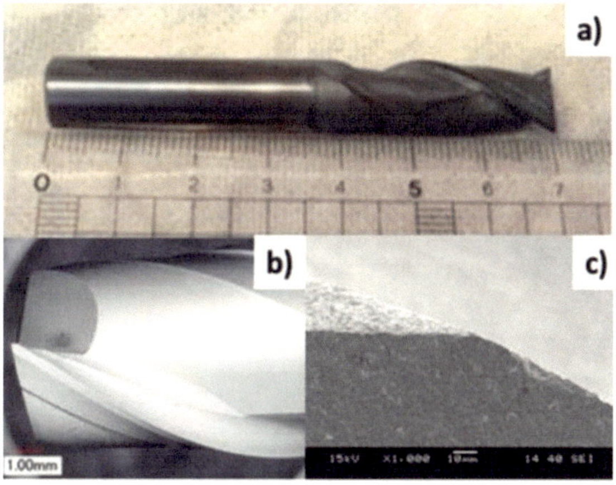

Figure 12. Ashing of CVD-diamond coating on the WC (Co) standard specimen under (100 V; −500 V; 45 Pa; 3.6 ks). (a) Surface of ashed specimen before cleansing, (b) Ashed tool edge, (c) SEM image of tool edge.

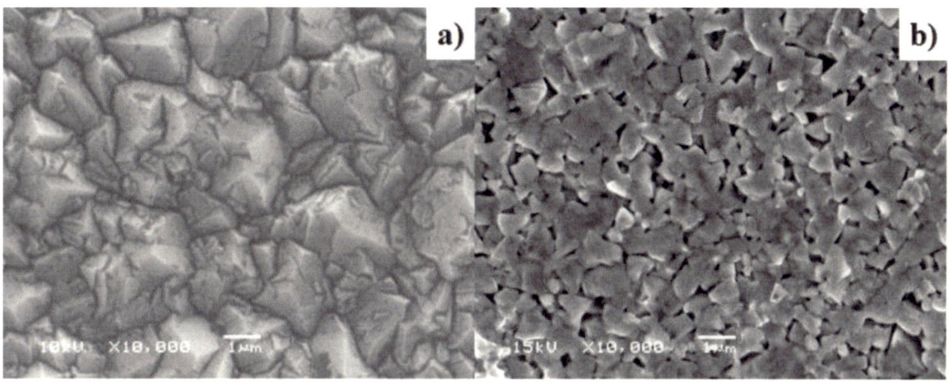

Figure 13. Comparison of microstructure on the diamond-coated tool surface before and after ashing under (100 V; −500 V; 45 Pa; 3.6 ks). (a) before ashing, and, (b) after ashing.

micro-pores are accommodated as a nucleation site for diamonds by selective removing the cobalt binders before CVD-coating. Hence, this microstructure in **Figure 13(b)** proves that starting WC (Co) substrate surface before coating is reproduced by the present ashing process. The Raman spectroscopy is also utilized to precisely check whether a residual debris of diamond is left on the ashed tool surface. **Figure 14** compares the Raman spectra before and after ashing. The presence of diamond film is identified by the Raman peak around 1330 cm^{-1} as shown in **Figure 14**. No peaks are detected at the vicinity of 1330 cm^{-1} in **Figure 14** after ashing; no diamond residuals are left on the tool surface after ashing.

3.4.2. Small-scaled end-milling tools

The small sized tools, as shown in **Figure 8**, were posttreated by plasma oxidation under the optimum condition in the above, or, (100 V; −500 V; 45 Pa; 3.6 ks). High resolution optical

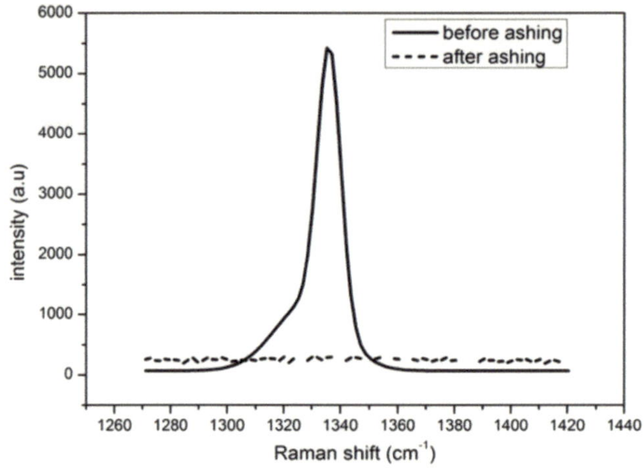

Figure 14. Comparison of the Raman spectra for the diamond-coated tool surface before and after ashing under (100 V; −500 V; 45 Pa; 3.6 ks).

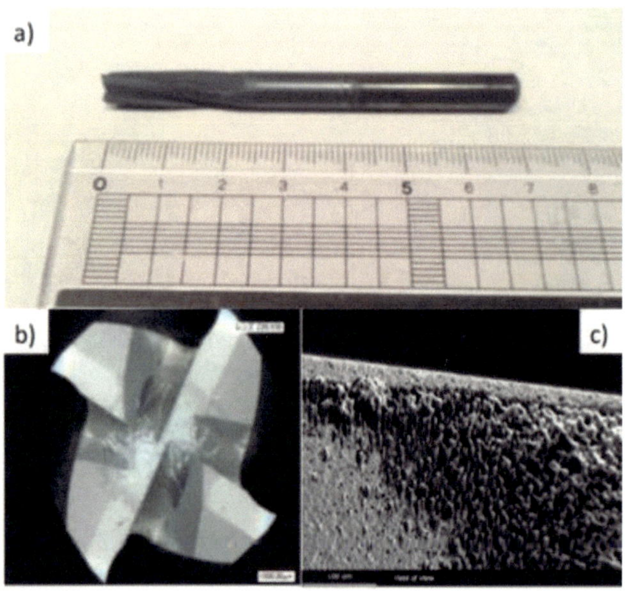

Figure 15. Ashing of CVD-diamond coating on the WC (Co) small-scaled specimen under (250 V; −500 V; 45 Pa; 3.6 ks). (a) Surface of ashed specimen before cleansing, (b) Ashed tool edge, (c) SEM image of tool edge.

microscopy and SEM are also utilized to make close observation on the ashed surfaces of small sized tools. As shown in **Figure 15(a)**, the outlook color of ashed tooth surfaces is gray in the similar manner to **Figure 12(a)**. **Figure 15(b)** shows the top tooth surfaces of ashed tools. No diamond films are left on the surfaces. These prove that most of diamond films coated on the small sized tools are also homogeneously removed just in correspondence to **Figure 12(b)** for the standard tool. After close observation on the whole tooth surfaces by SEM, a cold spot with the residual diamonds was detected on the rake face or the gash of tools. A typical SEM image is shown in **Figure 15(c)**. The diamond films at the upper edge and on the left-hand surface are almost completely ashed away; the diamond residuals are left as a dendritic column with the average diameter of 5–10 μm.

This difference in ashing behavior between two specimens comes from the optimality in oxygen plasma sheath distribution surrounding the tool surface. The plasma diagnosis is also employed to describe this effect of the plasma state on the ashing behavior.

3.4.3. In situ plasma diagnosis

EOS works to monitor the temporal variation of oxygen plasma state during the ashing process. Since the carbon in diamond reacts with oxygen atom, a peak for resultant of CO by the reaction, C (in diamond) + O (from O* and O^+ in palsmas) -> CO, is detected by EOS together with oxygen-related peaks. Hence, in this in situ EOS-diagnosis, oxygen peak at 470 nm and CO-peak at 288 nm are monitored to describe the plasma reaction for ashing.

Figure 16 depicts a time evolution of measured O- and CO-peak intensities by this in situ EOS. O-peak intensity is constant during ashing process; a constant oxygen atom flux is supplied to hot spots of plasma reaction. CO-peak intensity is enhanced in the early stage in ashing; it exponentially decays from t = 1000 s to be lower level and nearly minimum for t = 4000 s. This implies that homogeneous ashing process retards by itself with a constant time exponent. Hence, this exponential decay of CO-peak intensity with time proves that the diamond coating is homogeneously ashed away.

This time transient of CO-peak intensity is employed to describe the difference in ashing behavior between two diamond-coated tools. When using the standard tool, two CO-peak intensities, monitored at λ = 287 and 312 nm, exponentially decrease with time as shown in **Figure 17(a)**. Just as seen in **Figure 16**, the diamond film is expected to be homogeneously ashed away from this in situ diagnosis. This homogeneous ashing has been proven by precise observation on the posttreated tool surfaces as shown in **Figures 12–14**. On the other

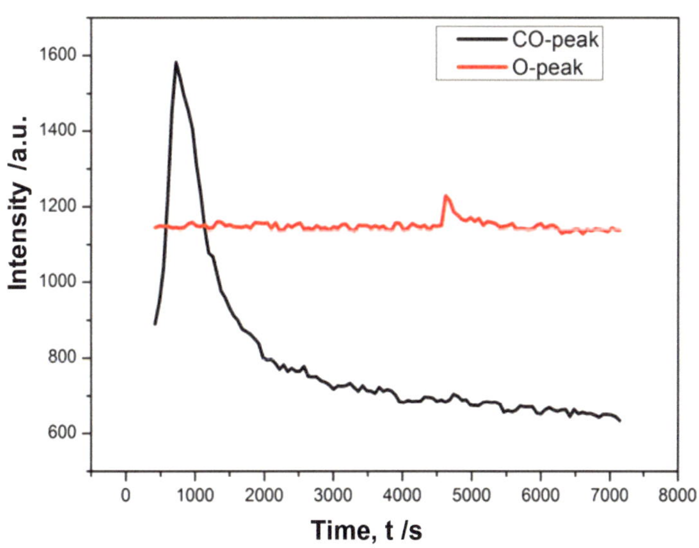

Figure 16. Temporal evolution of O- and CO-peak intensities, measured by the in situ EOS.

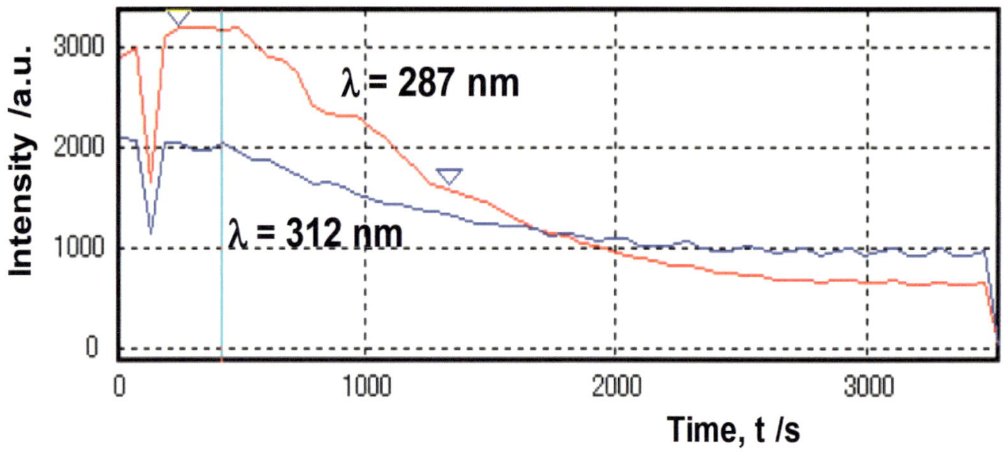

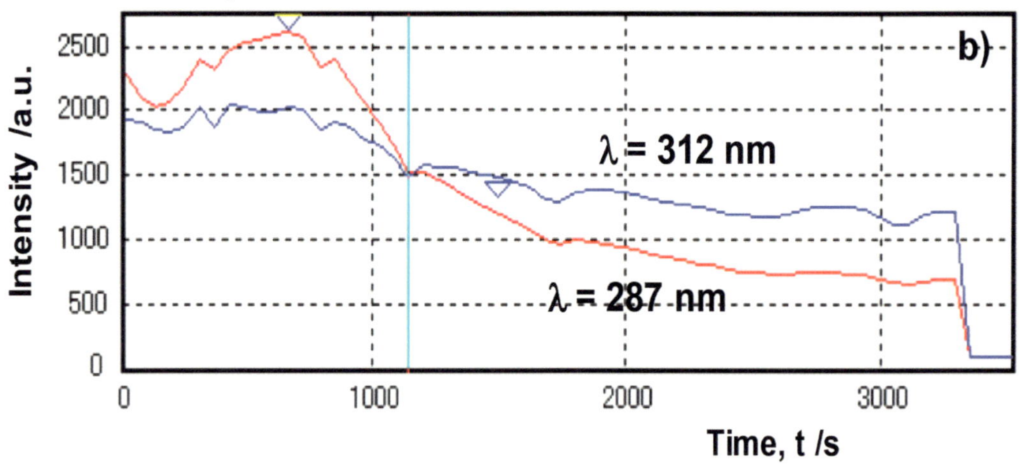

Figure 17. Time transients of CO-peak intensities at λ = 287 and 312 nm by the EOS in situ monitoring during ashing process.

hand, when using the small-sized tool, this time transient changes by itself as depicted in **Figure 17(b)**. The up-and-down deviation of CO-peak intensity reveals that ashing reaction localizes in hot spot at one surface but in cold spot at other surfaces. This heterogeneous ashing process results in a residual diamond film at the cold spots such as a small part on the rake edges of tools as shown in **Figure 15(c)**.

3.5. Summary

A diamond film coated onto a single WC (Co) tool substrate in a standard size was successfully ashed away without residuals and with much less damage at the tool tooth-edges. Due to this perfect ashing of diamond films, an original WC (Co) substrate surface was reproduced to have a decobalted WC (Co) microstructure. This posttreated surface is easy to be diamond-recoated for reuse of tools. The duration time for this ashing is only 3.6 ks, much faster than the commercial operation time by 20–30 hours experienced in several tooling companies. The whole ashing process can be in situ controlled by monitoring the CO-peak intensities at the specified wave length through EOS. A homogeneous, perfect ashing is guaranteed by an exponential decay of CO-peak intensity with time in this quantitative diagnosis. The tool teeth-edge geometry has significant effect on the formation of plasma sheaths on the tool surfaces. Inhomogeneous ashing can be also detected by the in situ plasma diagnosis; plasma oxidation conditions must be tailored for each tool geometry in practice.

4. Discussion

How to design this ashing process in practice and how to apply this posttreatment to industries are discussed in this session.

4.1. Plasma processing design

Theoretical plasma processing design is often needed to construct a new device to control a plasma configuration in chamber and to understand the experimental results [29]. Some studies have just started to build up this theoretical model. Among them, for an example, the Boltzmann equation was solved to obtain electron transport coefficients and rate coefficients for fluid models [30].

In the present study, a multi-disciplinary program is used to build up a theoretical model of plasma processing and to simulate the effect of pressure on the plasma density [11]. Although the basic data for activation process is still limited, one-dimensional model is proposed to describe the argon plasma state in the function of plasma parameters such as RF-voltage, DC-bias, and pressure. **Figure 18(a)** illustrates this one-dimensional model of argon plasma. Ionization of argon is expected to concentrate at the vicinity of tool surface in **Figure 9**. As depicted in **Figure 18(b)**, a hot spot with high electron temperature or the plasma sheath is formed only at the vicinity of cathode table. Under this geometric configuration, the electron density distribution is calculated at each specified pressure. **Figure 19** depicts a variation of electron density distribution with increasing the argon pressure. This enhancement

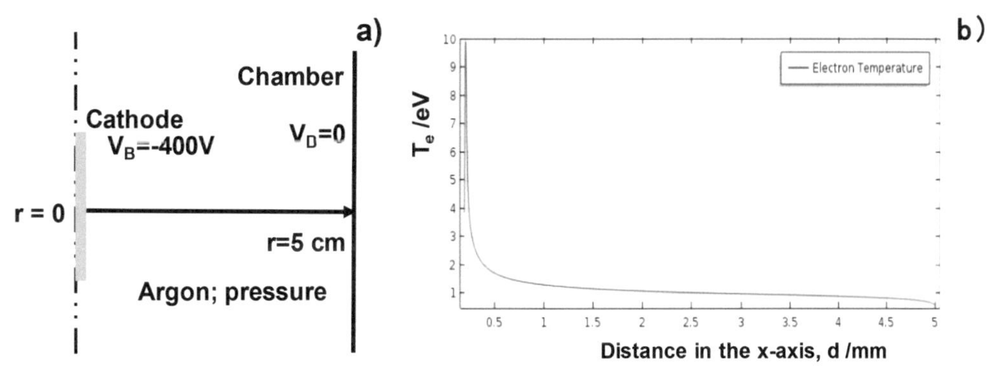

Figure 18. One dimensional model of argon plasma by using the multi-disciplinary equations. (a) Theoretical model of argon plasma, and, (b) decay of calculated electron temperature.

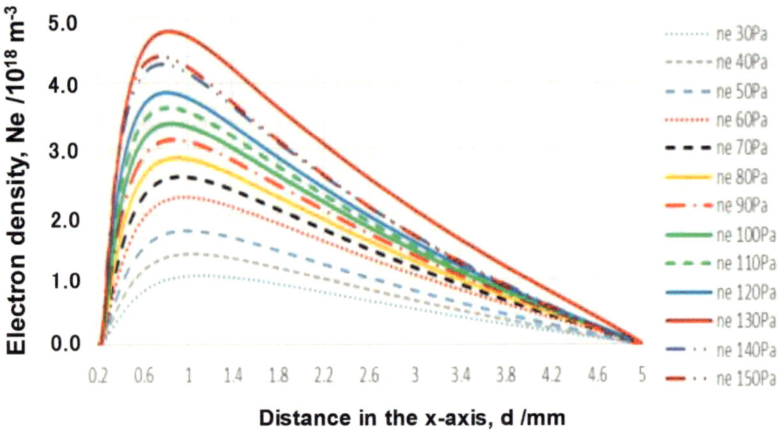

Figure 19. Enhancement of electron density distributions with increasing the argon pressure.

of electron density just near the cathode implies that electrons are activated in the plasma sheath. This theoretical model can be extended to describe the plasma oxidation process even by two dimensional model.

4.2. Down-sizing of plasma oxidation processing

Surface and interface area ratio to volume increases by down-sizing or miniaturing the dimensional size of products and elements. Surface or interface quality and functionality must be improved by surface- and interface -treatment and modification. The present plasma oxidation system has to be also down-sized to make plasma processing onto the surfaces and interfaces. **Figure 20(a)** illustrates a typical design of miniature plasma oxidation unit. RF-plasma is confined by using the RF-coils to form a plasma sheath around a small-sized tool above the insulator. The population of activated species in this plasma can be also controlled by the DC-bias. A shown in **Figure 20(b)**, a small-sized tool is subjected to high intensity plasmas in the designated space [18, 31].

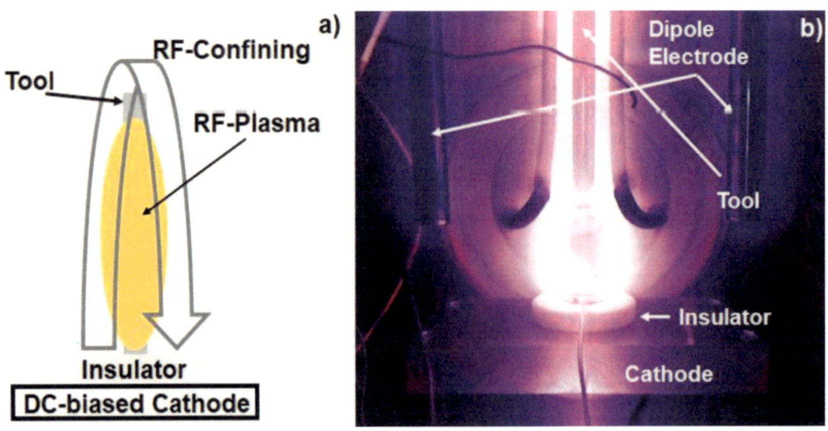

Figure 20. Down-sizing of an ashing unit for plasma oxidation of miniature parts and tools.

4.3. Plasma oxidation processing for industries

A single diamond-coated WC (Co) tool can be ashed away without residuals and tool edge damage by using the present plasma ashing system in **Figure 9**. Most of tooling companies are demanded to deal with hundreds of used diamond-coated tools in a day per a processing machine. The above plasma oxidation ashing system must be modified to make several or tens of works simultaneously even by using a single machine. The present system in **Figures 1** and **9** with use of the hollow cathode device is extended in the design concept to make simultaneous plasma oxidation for two diamond-coated tools [31, 32]. **Figure 21** depicts a developed simultaneous ashing system. The chamber is automatically opened for experimental set up of the diamond-coated tools, capped by the hollow cathode tubes.

An inserted photo in **Figure 21** shows that a pair of diamond-coated tool capped with the hollow cathode is subjected to the plasma oxidation. Only the inside of two hollows shines white; outside of hollows are in black. This reveals that oxygen plasmas are simultaneously confined into two hollows with high density and that ion density is much very low outside the hollows. This generation of oxygen plasmas is not limited to two hollows but extended to multi-ashing process to deal with tens of used diamond-coated tools by a single shot.

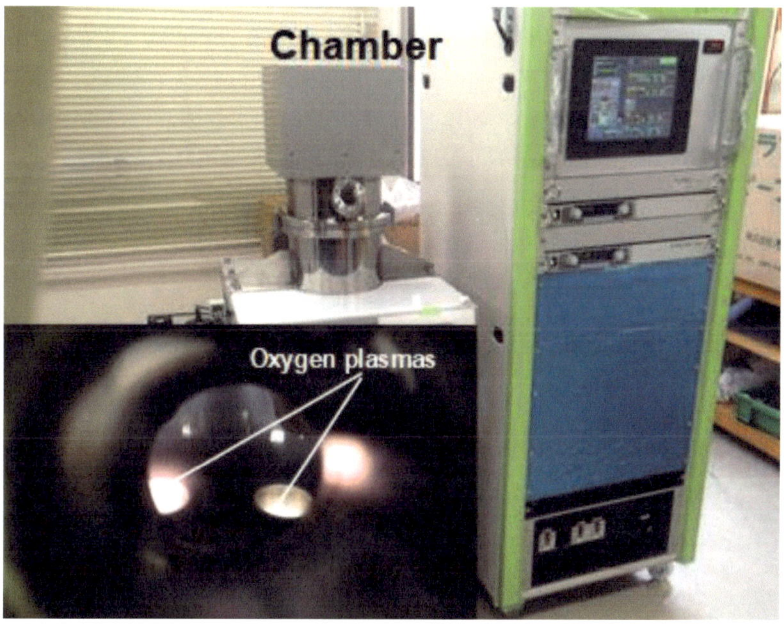

Figure 21. Simultaneous ashing system of two diamond-coated cutting tools toward industrial demonstration.

5. Conclusion

High density plasma oxidation ashing provides an effective processing to remove the used diamond films coated on the WC (Co) tool for recoating and reuse the mother tool substrate of WC

(Co). Different from the conventional plasma ashing processes in commercial, the diamond films are removed away without residuals even on the backsides of tool teeth and with less loss of teeth edges. In particular, the shrinkage in the tooth edge is more reduced down to 1.1 μm under the optimum ashing conditions than 5 μm in the commercial processes. In addition, the original microstructure with decobalted WC grains is reproduced by the present method to improve the engineering compatibility for nucleation and growth of diamond layers during the recoating process.

The emissive light optical spectroscopy as well as the Langmuir probe, equipped with the present system, work as a sensing medium to make quantitative diagnosis on the oxygen plasma state for optimization and to make in situ monitoring on the plasma oxidation reaction during the ashing process.

There might be several steps to extend this technology toward industrial applications. The carbon-base films including the diamond films are often utilized as a substrate of MEMS and miniature media; the small-scaled plasma ashing device is expected to work as a cutting and drilling tool to reshape that substrate and to modify its surface and interface. A miniatured oxygen plasma generator is just on the research to make demonstration tests. In practical operations of ashing processes in tooling industries, a stack of used diamond coated WC (Co) tools is processed by 100–1000 per a day. Considering that the tact time for present ashing is only 3.6 ks or 1 hour, tens of tools must be posttreated. The simultaneous plasma oxidation treatment is expected to be a candidate processing to make ashing of used diamond-coated tools in the industrial scale.

With aid of the theoretical model and simulation, the present plasma oxidation process can be extended to etching, polishing and finishing of various carbon base films, substrates and solid products by using the tailored devices to each application.

Acknowledgements

The author would like to express his gratitude to Dr. E.E. Yunata (SIT), Mr. K. Yamauchi (SIT), Ms. K. Kaminaga (SIT), and the late Mr. Y. Sugita (YS-Electric Industry, Co. Ltd.) for their help in experiments. This study was financially supported in part by the Grand-in-Aid from MEXT and by the supporting industry project from METI, Japan, respectively.

Conflict of interest

No conflict of interest was declared.

Author details

Tatsuhiko Aizawa

Address all correspondence to: taizawa@sic.shibaura-it.ac.jp

Surface Engineering and Design Laboratory, Shibaura Institute of Technology, Tokyo, Japan

References

[1] Hasegawa R. Cutting tools for aircraft and application. Journal of the Japan Society for Precision Engineering. 2009;**75**(8):953-957

[2] Bhushan B. Nano-tribology and nano-mechanics of MEMS/NEMS and bioMEMS/bioNEMS material and device. Microelectronic Engineering. 2007;**84**:387-412

[3] Kohn E, Gluche P, et al. Diamond MEMS, a new emerging technology. Diamond and Related Materials. 1999;**8**:934-940

[4] Kurita T et al. Advanced material processing with nano- and femto- second pulsed laser. International Journal of Machine Tools and Manufacture. 2008;**48**:220-227

[5] Allen DM et al. Ion beam, focused ion beam, and plasma discharge machining. Manufacturing Technology. 2009;**58**:647-662

[6] Syrkin A et al. Reactive ion etching of 6H-SiC in an ECR plasma of CF4 -O2 mixture using both Ni and Al mask. Materials Science and Engineering. 1997;**B46**:374-378

[7] Aizawa T, Sugita Y. High-density plasma technology for etching and ashing of carbon materials. Research Reports SIT. 2011;**55**(2):13-22

[8] Aizawa T, Mizushima K, Redationo TN, Yang M. Micro-imprinting onto DLC and CNT coatings via high density oxygen plasma etching. In: Proc. 8th ICOMM Conference (Canada, Victoria); 2013. pp. 459-466

[9] Yamauchi K, Yunata EE, Aizawa T. High density oxygen plasma ashing of used CVD diamond coating for recycling of WC (Co) tools. In: Proc. 4th IFMM & IFBF (2015, Toyama); pp. 233-236

[10] Yunata EE, Aizawa T. Micro-texturing into DLC/diamond coated molds and dies via high density oxygen plasma etching. Manufacturing Review. 2015;**2**:1-8

[11] Yunata EE. Characterization and application of hollow cathode oxygen plasma [PhD. thesis]. Shibaura Institute of Technology; 2016

[12] Aizawa T. Micro-patterning onto diamond like carbon coating via RF-DC oxygen plasma etching. In: Proc. 8th SEATUC Conference (Vietnam, Hanoi); 2011. pp. 425-428

[13] Aizawa T, Fukuda T. Micro-texturing onto carbon-based coatings via oxygen plasma etching. Research Reports SIT. 2012;**56**(29):66-73

[14] Suenaga R, Yunata EE, Aizawa T. Quantitative plasma diagnosis on high density RF-DC plasmas for surface processing. In: Proc. 7th SEATUC Conference OS6 19; 2013. pp. 1-6

[15] Cvelbar U et al. Inductively coupled RF oxygen plasma characterization by optical emission spectroscopy. Vacuum. 2008;**82**:224-227

[16] Aizawa T, Fukuda T. Oxygen plasma etching of diamond-like carbon coated mold-die for micro-texturing. Surface and Coating Technology. 2013;**215**:364-368

[17] Aizawa T, Morita H, Kurozumi S. Synthesis of films, coating system, coated products, dies and tools. In: Japanese Patent 2011-208140. 2011

[18] Aizawa T, Sugita Y. Distributed nitriding systems for surface treatment of miniature functional products. In: Proc. 10th 4M/ICOMM; 2015. pp. 449-453

[19] Bardos L. Radio frequency hollow cathodes for the plasma processing technology. Surface and Coating Technology. 1996;**46**:648-656

[20] Kitano A. The CFRP which supports the light-weighting of the plane. Journal of Chemical Education. 2011;**59**(4):226-229

[21] Aizawa T, Sugita Y. High density oxygen plasma ashing for recycling and reuse of DLC-coated tools and dies. Research Reports SIT. 2014;**58**(2):1-10

[22] A. Gilpin, Composite Drill; Tool Process Change Nets 450 Percent More Holes per Bit Life. http://www.diamondtc.com/media/tooling_production/index.html [21/4/2018]

[23] Aizawa T, Masaki E, Sugita Y. Oxygen plasma ashing of used DLC coating for reuse of milling and cutting tools. In: Proc. Int. Conf. Mater; Advanced Process Technology; 2011. pp. 15-20

[24] Aizawa T, Masaki E, Sugita Y. Complete ashing of used DLC coating for reuse of the end-milling tools. Manufacturing Letters. 2014;**2**:1-3

[25] Aizawa T, Masaki E, Moromoto E, Sugita Y. Recycling of DLC-coated tools for dry machining of aluminium alloys via oxygen plasma ashing. Mechanical Engineering Research. 2014;**4**(1):52-62

[26] Yamauchi K, Aizawa T. Optimization of diamond ashing process for recoating of CVD diamond coated tools. In: Proc. 8th AWMFT; 2015. pp. J16/1-J16/6

[27] Yunata EE, Yamauchi K, Aizawa T. High density plasma ashing of CVD diamond coated end-milling tools. In: Pro. 4th ASMP2015 (Lombok, Indonesia) 2015: CD-ROM

[28] Yamauchi K, AizawaT. High density plasma ashing of used diamond coated short-shank tools without damage to WC (Co) teeth. In: Proc. 11th ICOMM 6; 2015. pp. 1-6

[29] Mihailova D, Grozeva M, et al. Theoretical and experimental studies of the plasma processes in hollow cathode discharge lasers. In: Proc. 28th ICPIG, 98; 2007. pp. 533-536

[30] Hagelaar GJM, Pitchford LC. Solving the Boltzmann equation to obtain electron transport coefficients and rate coefficients for fluid models. Plasma Sources Science and Technology. 2005;**14**:722-733

[31] Aizawa T. Development of micro-manufacturing by controlled plasma technologies. Journal of Japan Society for Technology of Plasticity. 2017;**58**(12):1064-1068

[32] Aizawa T. Future prospects and advanced steps of die and mold technologies. Journal of Japan Society for Technology of Plasticity. 2018;**59**(1):19-23

Section 3

CVD of 2D Materials

Chapter 5

Large-Area Synthesis and Growth Mechanism of Graphene by Chemical Vapor Deposition

Chen Wang, Kizhanipuram Vinodgopal and
Gui-Ping Dai

Additional information is available at the end of the chapter

http://dx.doi.org/10.5772/intechopen.79959

Abstract

There has been continuous progress in the development of different synthesis methods to readily produce graphene at a lower cost. Compared with the other methods, chemical vapor deposition (CVD) is an effective and powerful method of producing graphene and has attracted increased attention during the last decade. In this way, we can obtain good uniformity with a multitude of domains, excellent quality, and large scale of the produced graphene. Meanwhile, it is also helping for large-area synthesis of single-crystal graphene. In the CVD method, precursors are typically absorbed on the surface followed by pyrolytic decomposition, which leads to the generation of absorption sites on the surface and promotes the growth of continuous thin films.

Keywords: chemical vapor deposition, graphene, large-scale synthesis, growth mechanism

1. Introduction

Graphene (shown in **Figure 1**), as a versatile two-dimensional material [1], has attracted great interest in the research community owing to its high mechanical stiffness [2], great flexibility [3], and stable chemical properties [4–6]. Based on the properties of graphene and its nanoscale thickness, it has been selected as a material for various applications in electrical, chemical, mechanical, optical, and catalysis industries [7–11]. Hence, it is very critical to propose methods for preparing high-quality graphene film.

Compared with other methods, such as mechanical exfoliation of graphite [12–14], liquid-phase exfoliation [15, 16], and reduction of graphene oxide (GO) [17, 18], chemical vapor deposition

Figure 1. The schematic diagram of graphene.

(CVD) is regarded as the most promising way for large-scale graphene production at a large scale with low defects, good uniformity, and controlled number of graphene layers, which has attracted intense research attention during the last decades [19–22]. CVD involves the activation of gaseous reactants and the subsequent chemical reaction, followed by the formation of a stable solid deposit over a suitable substrate such as Ni [6, 23], Cu [5, 24, 25], Fe [26], Pt [27, 28], or their alloys [29, 30]. As early as 1969, Robertson et al. [31] discovered that chemical vapor deposition in the presence of methane produces a layer of graphite on some transition metal surfaces. However, even if the prepared graphene displays excellent performance, it cannot be applied to the related devices due to the limitations of deposition on the metal surface. The advent of poly(methyl methacrylate)-mediated nanotransfer printing technology [32] enabled the successful transfer of graphene from the growth substrate to any other substrate, and it also completely changed the status of chemical vapor deposition methods in the preparation method of graphene.

It is worth mentioning that, compared to exfoliated graphene obtained from natural graphite, the CVD graphene still shows a lower carrier mobility ranging from 100 to 1000 $cm^2/(V \cdot s)$, due to the presence of growth defects and boundaries [33]. Therefore, a feasible route to improve graphene film quality lies in increasing the domain size that enables to avoid the side effect of graphene domain boundaries.

Now, large-area graphene synthesized have generally been single crystalline, and related research is aimed to control the domain size, defects, number of graphene layers, and so on. With regard to desired graphene, there are many effect parameters including hydrogen, oxygen, gas flow rate, and residence time.

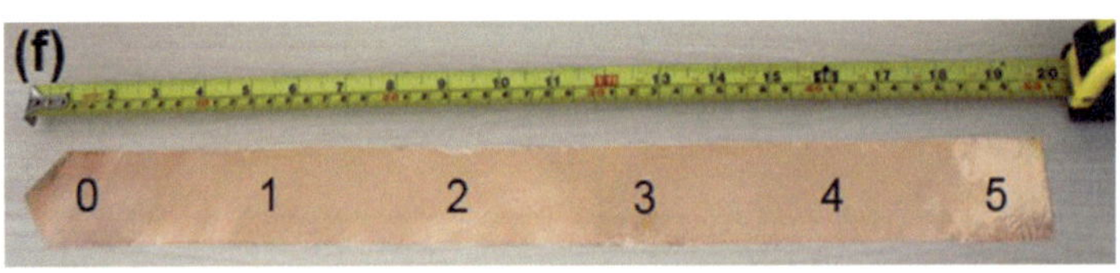

Figure 2. Image of meter-size single-crystal graphene. Reproduced with permission of ref. [34].

http://dx.doi.org/10.5772/intechopen.79959

In addition, great success has been achieved in the fabrication of large-area, single-crystalline graphene. Different teams have published reports on the single crystals. We also summarize these experimental results.

There are already many comprehensive articles published about increasing the size of graphene. Among these articles, a publication by Xu et al. [34] reports the fabrication of meter-size single-crystal graphene. The authors developed an innovative method to achieve the ultrafast epitaxial growth of meter-sized single-crystal graphene on a commercially available Cu foil (shown in **Figure 2**).

In this chapter, we talk about the preparation of high-quality graphene in recent years, effect of experimental parameters, challenges of preparation of high-quality graphene, and timeline of graphene production. Finally, development trend and application prospect of CVD graphene are described.

2. Preparation of graphene

Great effort has been given to the synthesis of graphene with controlled size, morphology, edge structures, and layer numbers [22, 35–40]. Studies into the dynamics of CVD graphene growth on copper foils have chiefly focused on monolayer films that develop under vacuum conditions. Up to now, the preparation methods of graphene are usually divided into two ways: physical and chemical (shown in **Figure 3**). With respect to physical method, mechanical exfoliation [12–14] and liquid-phase exfoliation [15, 16] are always used. With respect to chemical methods, they include CVD, reduction of graphene oxide (GO) [17, 18], epitaxial growth on various substrates.

2.1. Mechanical exfoliation

The mechanical exfoliation method can provide the highest quality graphene sheets, which strips the graphene layers from the surface of graphite crystals by mechanical force. It is not suitable for mass production, however, because of the difficulty in controlling the number of layers and the size of sheets.

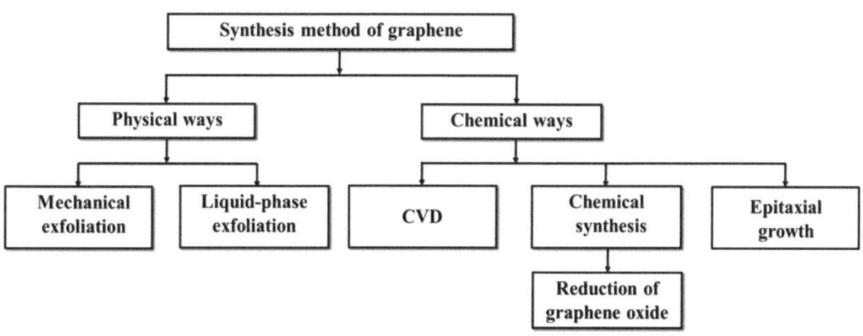

Figure 3. Classification of graphene synthesis techniques.

2.2. Liquid-phase exfoliation

Liquid-phase exfoliation cannot produce graphene films with high electronic quality, where scatter graphite to specific solvents or surfactants, and strip the single layer or multilayer graphene through ultrasonic energy to get graphene dispersions [15, 41].

2.3. Reduced graphene oxide (RGO)

With regard to RGO methods, first, the graphite is oxidized by treatment with $KMnO_4$, $NaNO_3$, and so on. Second, GO prepared from flake graphite can be readily dispersed in water and has been used on a large scale for preparing large graphitic films, but it would result defects in graphene [42].

2.4. Chemical vapor deposition

In comparison, CVD is an effective and powerful method of producing graphene. CVD processes typically involve two steps: the activation of gaseous reactants and the chemical reaction of forming a stable solid deposit over a suitable substrate [43, 44]. The deposition process consists of two types of reactions: homogeneous gas-phase reactions, occurring in the gas phase, and heterogeneous chemical reactions that occur on a heated surface. Therefore, CVD method is the use of carbon source (gas, liquid, and solid) decomposing in the high-temperature reaction zone, the release of carbon atoms in the metal substrate, eventually it forms a continuous graphene membrane. And many researchers use CVD method to prepare the high-quality and large-area graphene, as Li et al. [45] pointed (shown in **Figure 4**). Subsequently, CVD can be divided into two types: (I) thermal CVD and (II) plasma-enhanced CVD.

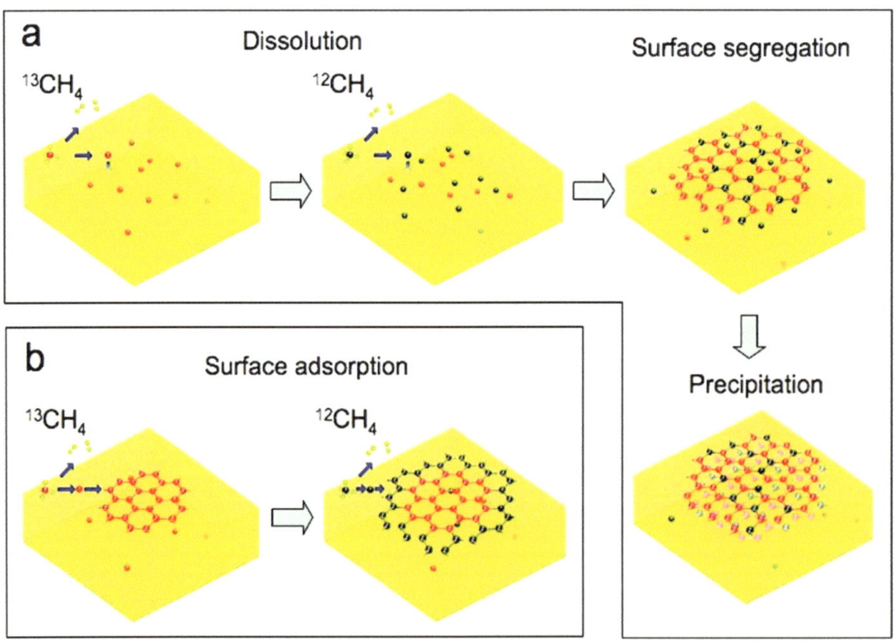

Figure 4. Schematic diagram of growth mechanism of graphene on the metal substrate. Reproduced with permission of ref. [45].

2.5. Epitaxial growth on substrates

Obviously, the term "epitaxial growth" means ordered or arranged atomic growth in single crystalline manner over a single-crystalline substrate. When the substrate deposited materials are the same as the substrate, it is called homoepitaxial growth. Oppositely, it is defined as hetero epitaxial growth if substrate differs from the deposited materials. And epitaxial growth of graphite on semiconducting substrates (SiC) was known to us since 1975 [46]. Berger's group successfully synthesized epitaxial graphene sheet [47] by using Si terminated face of single-crystal 6H-SiC by thermal desorption of Si. To summarize, although it is still an expensive approach, this is a promising method and possibly be scalable commercially.

3. CVD synthesis mechanism and the choice of carbon source

Chemical vapor deposition has been used to prepare inorganic materials in recent years. Besides, it also can be applied to produce graphene. CVD is the process of using gaseous matter to react on a solid surface and form solid deposits. The thermal decomposition and thermal synthesis of the deposition techniques are carried out at high temperature. As for the synthesis of graphene, three steps are generally required:

1. a volatile precursor,
2. transport of volatile matter to the precipitation area, and
3. chemical reaction on the substrate to produce a solid substance.

Two phenomenological growth mechanisms have been proposed based on an isotope labeling experiments [12] to understand the graphene growth. It is very easy to form a metal-carbide phase due to the strong interaction between carbon and metal [45, 48–50], following the precipitated growth mechanism. During the subsequent cooling process, the pyrolytic carbon atoms from the carbon sources firstly dissolve into the catalyst and then deposit at the metal surface to form graphene layers. On the other hand, the pyrolytic carbon atoms could only diffuse on the catalyst surface as the cases are shown with Cu and Au when the interaction between metal and carbon is fairly weak [22, 51], where both the nucleation and growth of graphene are dominated by the surface diffusion of the decomposed carbon atoms. Multi-layered graphene film formation normally follows the sedimentary growth mechanism and the number of graphene layers can be controlled by adjusting the dissolved C atoms in the metals or the thickness and composition of the substrate or by slowing down the cooling rate. Obviously, only single-layer graphene can be formed in the diffusive growth because the feedstock cannot get access to the graphene covered area of catalyst surface.

Meantime, SLG (shown in **Figure 5**) has been synthesized from solid and liquid hydrocarbon sources at different temperature (400–1000°C). Li et al. [52] adopted polymethyl-methacrylate (PMMA) and polystyrene as solid hydrocarbon and benzene as a liquid hydrocarbon source to grow graphene at temperatures lower than 1000°C on copper foil. Jang et al. [53] tried to remove remaining oxidizing impurities (oxygen-free APCVD, which is shown in **Figure 6**) before graphene deposition by using pumping and purging. This method leads to continuous SLG with full coverage of the substrate.

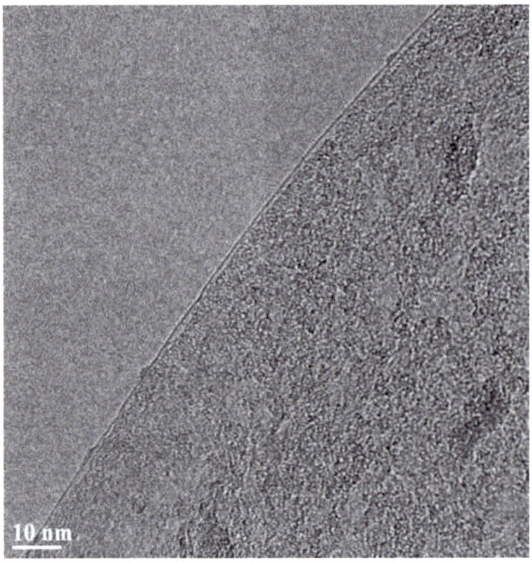

Figure 5. HRTEM image of single-layer graphene.

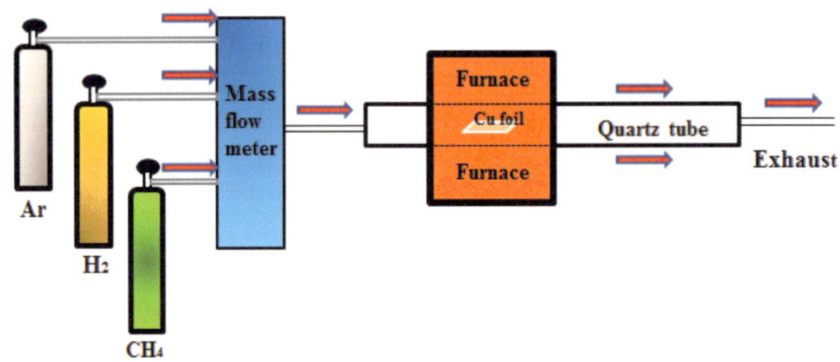

Figure 6. The schematic diagram of APCVD. Reproduced with permission of ref. [54].

As for the multilayer graphene (MLG), the high solubility of carbon in catalysts results in the dissolution of carbon into the bulk at high temperatures, leading to multilayer graphene (MLG) during the cooling process [33, 54–56].

4. Toward high-quality graphene: effect of experimental parameters

4.1. Choice of metal base material

Up to now, a large number of research have been done about the growth of graphene on the metal surfaces. CVD graphene growth is strongly dependent on the catalyst. Due to the different catalytic activity and atomic packing, the growth behavior of graphene on catalyst surfaces, such as quality, continuity, and layer number, is different in each case. The catalyst plays an important role in the CVD process. We generally use Cu [5, 24, 25], Ni [6, 23], Pd [49], Ru [57], and Ir [58], as the metal base material for the preparation of graphene. As we know, different metal base materials have a different effect on the preparation of graphene, so the metal substrate is the key factor to determine the growth of graphene.

The mechanism of growth of graphene by chemical vapor deposition in copper and nickel substrate is different. Studies have shown that the growth of graphene on other metal substrates can also be basically classified into the growth mechanisms of copper and nickel substrates. On a metal substrate with high soluble carbon content, such as nickel, the carbon atoms produced by the carbon source cracking are infiltrated into the metal matrix at high temperature. When the temperature drops, the core is precipitated from the inside, then the graphene is formed. It is difficult to control the precipitation of carbon atoms during the cooling process because the solubility of carbon atoms in nickel increases with temperature, so it has a negative effect on the preparation of graphene. In the case of copper, the solubility of carbon atoms in copper is relatively low at high temperature; therefore, compared with other catalysts, Cu is currently the most widely used catalyst material [22, 45].

4.2. Effect of hydrogen

Hydrogen is often involved in CVD graphene synthesis. Thus, hydrogen has a great impact on graphene reducing substrate surface contamination and defects in the annealing process [59], controlling the graphene domain shape [40, 60, 61], nucleation [62], and layer number [63]. And excess hydrogen could etch graphene to destroy the integrity of the lattice and make the quality of graphene worse [64]. It also affects the adsorption, stability, thickness [65–67], population of active species on the catalyst surface [68, 69] and the morphology [70–79] of the grown graphene. Luo et al. [72] determined the effect of hydrogen on the shape and orientation of graphene using DFT calculation (shown in **Figure 7**). High hydrogen concentration is required to synthesize large single-crystal monolayer graphene domains [40, 80]. Besides the low carbon concentration, etching by hydrogen is also an obstacle to short-time graphene growth [81].

4.3. Effect of oxygen

It is known that oxygen plays an important role in each step of graphene synthesis, especially on copper. The effect of surface oxygen or pre-adsorbed oxygen on methane decomposition on Cu and that on the transition metal surface such as Ni and Pd is different [33].

The strong binding of active carbon species with catalyst surface is crucial for nucleation of graphene. The adsorption of active species on the catalyst surface becomes higher in the presence of oxygen. Hao et al. [82] found that the graphene edge with H-termination is better than a bare edge on Cu. In addition, oxygen plays a vital role in reducing the nucleation density not only in the bulk but also on the surface [83].For example, oxygen could remove the unwanted carbon from Cu bulk, [84] which results in unintentional nucleation.

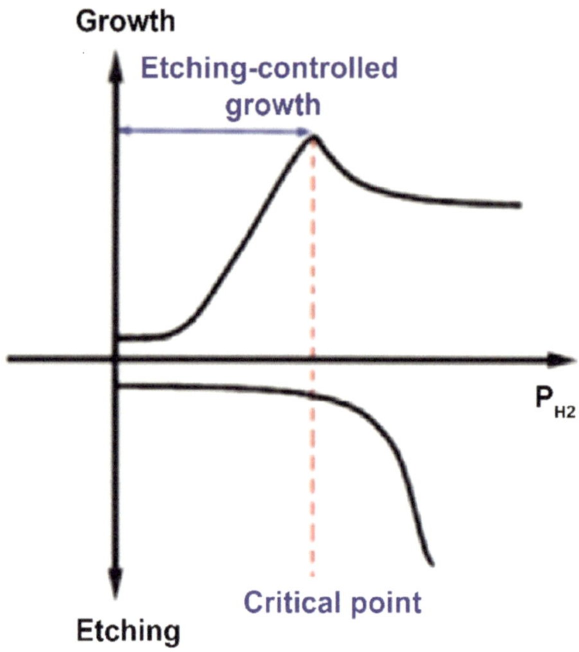

Figure 7. Schematic diagram of Etching-controlled growth mechanism of CVD. Reproduced with permission of ref. [71].

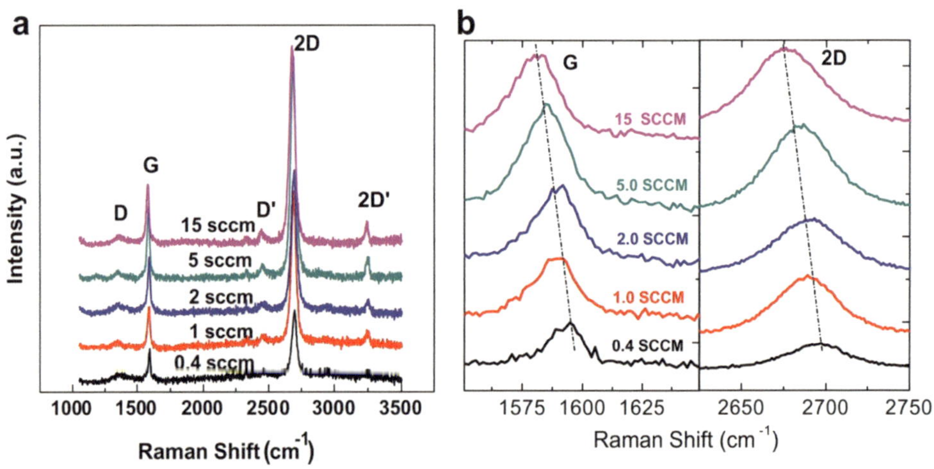

Figure 8. Raman spectra (a) and Gand 2D peaks (b) of the graphene films grown under various methane fluxes. Reproduced with permission of ref. [85].

4.4. Effect of gas flow rate

The gas flow has been considered as a crucial factor during graphene synthesis. Optimizing the flow rate of CH_4 and H_2 could improve the quality of graphene. Therefore, Li et al. [22] described the CVD growth of graphene single crystals up to 0.5 mm in size in a quasi-static flow regime, using a copper enclosure in LPCVD. Meantime, Wang et al. [85] have used a H_2

http://dx.doi.org/10.5772/intechopen.79959

flow rate of 20 sccm adopt, at 80 Pa, and 1000°C, in their CVD method for graphene growth. The methane flow was varied systematically between 0.4 and 15 sccm. The graphene was characterized by Raman spectroscopy, and the results were shown in **Figure 8**. The study found that the ratio of I_G/I_{2D} peak ratio did not increase with the increase of methane flow and is less than 0.5. This indicates the formation of single layer graphene and no dependence on methane flow rate. They reported that the methane flow has a significant effect on the structure of graphene and determined the optimal flow rate of methane as 5 sccm. At 0 sccm, I_D/I_{2D} peak intensity ratio has a minimum value (0.0395) yielding graphene with the least defects.

5. Timeline of graphene production

Li et al. [22] synthesize large-area graphene grown by low-pressure chemical vapor deposition in copper-foil enclosures using methane as a precursor single crystal with dimensions of up to 0.5 mm on a side. While Hao et al. [82] enabled repeatable growth of centimeter-scale single-crystal graphene domains to attain 1 cm single-crystal graphene by controlling of surface oxygen. Correspondingly, Wu et al. [29] prepare a similar to 1.5-inch-large graphene monolayer in 2.5 h by locally feeding carbon precursors to a desired position of a substrate composed of an optimized Cu-Ni alloy. Besides, Nguyen et al. [86] have fabricated the 6 × 3 cm^2 graphene film without grain boundaries on polished copper(111) foil by Seamless stitching of graphene domains. However, a foundation of the modern technology by Xu et al. [34] that using single-crystal silicon improve the growth of high-quality single-crystal graphene with diameters up to 12 inches or larger.

In 2004, micron-sized graphene was obtained by exfoliation, which was not scalable. In 2006, the scientists obtained graphene by thermal deposition of SiC, which is an expensive process. In 2009, researchers first used Cu-CVD method to get graphene, and subsequently, this CVD method was extended to using other metals as substrates.

The graphene can be transferred using a wet method [32]. The process involves (1) coating a protective layer (e.g., PMMA) on one side, (2) etching graphene on the other side, (3) etching away the copper and then rinse with water, (4) transferring to target substrate, and (5) removal of protective film, by typically using solvent to dissolve it. With the development of graphene production, there are many ways to transfer graphene, including R2R ([87–89],

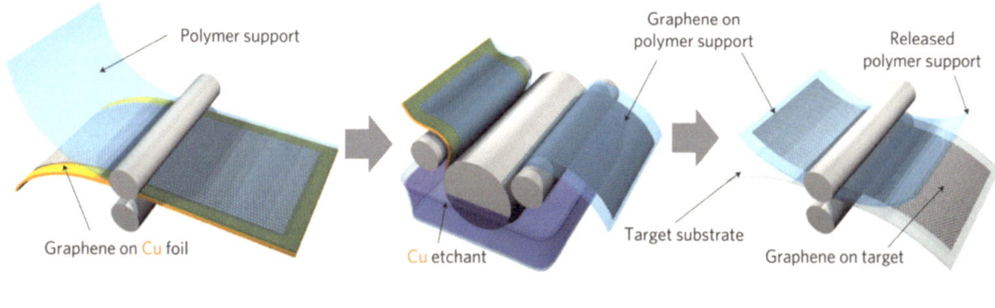

Figure 9. Schematic diagram of roll to roll production of graphene. Reproduced with permission of ref. [89].

shown in **Figure 9**) with thermal release tape, R2R with epoxy [90], and bubbling transfer [91]. These developments have led to much more controllable graphene synthesis [19, 33, 89].

6. Development trend and application prospect of CVD graphene

As for the graphene synthesis by CVD, it is beneficial to maximize the scale while maintaining the homogeneity of large-area films, as indicated by Bae et al. [89]. A creative approach to the CVD synthesis of ultra large-area graphene based on selective Joule heating was reported by Kobayashi et al. [92]. Since the first growth on copper foil a decade ago, inch-sized single-crystal graphene has been achieved. For most industrialized applications, large single-crystal graphene films are promising for top-down processing. In recent decades, the SCG island size has increased more than four orders of magnitude, from micrometers to inches. Xu et al. [34] have made large-area graphene film (5 cm × 50 cm) within 20 min, which is almost 99% ultra-highly oriented grains and is, to the best of our knowledge, the largest graphene area synthesized and demonstrated by far. The growth steps are described as follows: (1) manufacturing substrate with meter-sized single-crystal Cu(111) foil; (2) epitaxial growth of graphene grains on the Cu(111) surface; (3) seamless merging of graphene domains into a film with high-single crystallinity; and (4) the growth of graphene film in an ultrafast fashion.

At present, the process of preparing graphene by chemical vapor deposition has matured to the point of large-scale production and has been applied to electric ink, lithium battery additives, coatings and composite materials additives, touch screen, and other low-end applications [93].

For lithium battery applications, graphene as a conductive additive can also solve the contradiction between the high-voltage real ratio and the battery performance of the cathode materials such as lithium cobalt oxide [94]. Graphene has extremely high thermal conductivity, high thermal radiation coefficient, and high specific surface area, so that a surface coating of graphene as auxiliary heat dissipation has great potential for application [95]. Graphene can be used as the main filling material for new conductive ink and plastics [96], and it can also be mixed with nanometer silver powder into a new type of conductive adhesive or ink, and the resistivity and adhesion properties of conductive ink can be adjusted by changing the filling ratio of graphene. The two-dimensional structure of graphene allows for conductive thermal channels in graphene-based anticorrosive paint coating and forms a tight labyrinth of physical barriers to prevent corrosion related exposure. Graphene could be applied to improve performance on graphene capacitive screen [96]. Similarly, carbon nanowalls [97] described as a graphite nano-structure with edge structure and arranged perpendicularly to the matrix on a sheet of graphite have been synthesized by the CVD and is widely used in electrochemical devices, catalyst carriers, etc.

7. Summary and outlook

With excellent prosperities of graphene, there are tremendous amount of research on digging out its exceptional prospect in future applications. But challenges including the growth rate of graphene growth, defects, stacking order, and transfer still remain. Nevertheless, more and more scientific communities are making efforts to achieve the industrialization of graphene. Undoubtedly, the graphene industry is growing at a fast rate. In the next 5–10 years, the technology for graphene-based conductive additives, anticorrosive coatings, touch screen would be mature, along with the industrialization of high-end applications such as super capacitors, sensors, and electronic chips. But mono-crystalline graphene films are extremely difficult to prepare. At present, they can only be made in millimeter size, and they are still quite far away from the practical crystal size, which is difficult to break in the short term. We need to achieve a balance between the advantages and disadvantages of graphene in order to create a fast and healthy way for the sustainable development of graphene. Chemical vapor deposition (CVD) is the most promising preparation technique. The preparation by chemical vapor deposition has become a major method for preparing semiconductor, thin film materials, and industrial production of high-quality graphene materials in large quantity. Meantime, for the CVD method, considerable efforts have been exerted to get low cost on any arbitrary substrate, but it is still a long way to make it.

Author details

Chen Wang[1], Kizhanipuram Vinodgopal[2]* and Gui-Ping Dai[1,2,3]*

*Address all correspondence to: nanodai@gmail.com

1 School of Resources Environmental and Chemical Engineering, Nanchang University, Nanchang, Jiangxi, China

2 Department of Chemistry and Biochemistry, North Carolina Central University, Durham, NC, USA

3 Institute for Advanced Study, Nanchang University, Nanchang, Jiangxi, China

References

[1] Novoselov KS, Geim AK, Morozov SV, Jiang D, Katsnelson MI, Grigorieva IV, et al. Two-dimensional gas of massless Dirac fermions in graphene. Nature. 2005;438:197-200

[2] Liu Y, Xie B, Zhang Z, et al. Mechanical properties of graphene papers. Journal of the Mechanics & Physics of Solids. 2012;**60**(4):591-605

[3] Stöberl U, Wurstbauer U, Wegscheider W, et al. Morphology and flexibility of graphene and few-layer graphene on various substrates. Applied Physics Letters. 2008;**93**(5):19912

[4] Mišković-Stanković V, Jevremović I, Jung I, et al. Electrochemical study of corrosion behavior of graphene coatings on copper and aluminum in a chloride solution. Carbon. 2014;**75**(10):335-344

[5] Raman RKS, Banerjee PC, Lobo DE, et al. Protecting copper from electrochemical degradation by graphene coating. Carbon. 2012;**50**(11):4040-4045

[6] Nayak PK, Hsu CJ, Wang SC, et al. Graphene coated Ni films: A protective coating. Thin Solid Films. 2013;**529**(2):312-316

[7] Ma Y, Kim D, Jang H, et al. Characterization of low temperature graphene synthesis in inductively coupled plasma chemical vapor deposition process with optical emission spectroscopy. Journal of Nanoscience & Nanotechnology. 2014;**14**(12):9065

[8] Ning G, Xu C, Cao Y, et al. Chemical vapor deposition derived flexible graphene paper and its application as high performance anodes for lithium rechargeable batteries. Journal of Materials Chemistry A. 2012;**1**(2):408-414

[9] Chen Z, Ren W, Gao L, et al. Three-dimensional flexible and conductive interconnected graphene networks grown by chemical vapour deposition. Nature Materials. 2011;**10**(6):424

[10] Yuan GD, Zhang WJ, Yang Y, et al. Graphene sheets via microwave chemical vapor deposition. Chemical Physics Letters. 2009;**467**(4):361-364

[11] Zheng Q, Kim JK. Synthesis, structure, and properties of graphene and graphene oxide. Journal of Experimental Medicine. 2015;**23**(2):29-94

[12] Novoselov KS, Geim AK, Morozov SV, Jiang D, Zhang Y, Dubonos SV, et al. Electric field effect in atomically thin carbon films. Science. 2004;**306**:666-669

[13] Geim AK. Graphene: Status and prospects. Science. 2009;**324**:1530-1534

[14] Schwierz F. Graphene transistors. Nature Nanotechnology. 2010;**5**:487-496

[15] Hernandez Y, Nicolosi V, Lotya M, et al. High-yield production of graphene by liquid-phase exfoliation of graphite. Nature Nanotechnology. 2008;**3**(9):563-568

[16] Ciesielski A, Samori P. Graphene via sonication assisted liquid-phase exfoliation. Chemical Society Reviews. 2014;**43**:381-398

[17] Dikin DA, Stankovich S, Zimney EJ, Piner RD, Dommett GHB, Evmenenko G, et al. Preparation and characterization of graphene oxide paper. Nature. 2007;**448**:457-460

[18] Eda G, Fanchini G, Chhowalla M. Large-area ultrathin films of reduced graphene oxide as a transparent and flexible electronic material. Nature Nanotechnology. 2008;**3**:270-274

[19] Li M, Liu D, Wei D, et al. Controllable synthesis of graphene by plasma-enhanced chemical vapor deposition and its related applications. Advanced Science. 2016;**3**(11):1600003

[20] Qi Y, Eskelsen JR, Mazur U, et al. Fabrication of graphene with CuO islands by chemical vapor deposition. Langmuir the ACS Journal of Surfaces & Colloids. 2012;**28**(7):3489-3493

[21] Sun J, Lindvall N, Cole MT, et al. Low partial pressure chemical vapor deposition of graphene on copper. IEEE Transactions on Nanotechnology. 2012;**11**(2):255-260

[22] Li X, Cai W, An J, et al. Large-area synthesis of high-quality and uniform graphene films on copper foils. Science. 2009;**324**(5932):1312

[23] Kwon H-J et al. Synthesis of flake-like graphene from nickel-coated polyacrylonitrile polymer. Nanoscale Research Letters. 2014;**9**(1):618

[24] Kim M-S et al. Effect of copper surface pre-treatment on the properties of CVD grown graphene. AIP Advances. 2014;**4**(12):127107

[25] Gnanaprakasa TJ et al. The role of copper pretreatment on the morphology of graphene grown by chemical vapor deposition. Microelectronic Engineering. 2015;**131**:1-7

[26] Vinogradov NA et al. Formation and structure of graphene waves on Fe(110). Physical Review Letters.2012;**109**:026101

[27] Sutter P, Sadowski JT, Sutter E. Graphene on Pt(111): Growth and substrate interaction. Physical Review. 2009;**B80**:245411

[28] Feng X, Wu J, Bell AT, Salmeron M. An atomicscale view of the nucleation and growth of graphene islands on Pt surfaces. Journal of Physical Chemistry C. 2015;**119**:7124-7129

[29] Wu T et al. Fast growth of inch-sized single-crystalline graphene from a controlled single nucleus on Cu–Ni alloys. Nature Materials. 2015;**15**(1):43

[30] Liu X, Fu L, Liu N, Gao T, Zhang Y, Liao L. Segregation growth of graphene on Cu–Ni alloy for precise layer control. Journal of Physical Chemistry C. 2011;**115**:11976-11982

[31] Robertson SD. Graphite formation from low temperature pyrolysis of methane over some transition metal surfaces. Nature. 1969;**221**(5185):1044-1046

[32] Jiao L, Fan B, Xian X, et al. Creation of nanostructures with poly(methyl methacrylate)-mediated nanotransfer printing. Journal of the American Chemical Society. 2008;**130**(38): 12612-12613

[33] Addou R et al. Monolayer graphene growth on Ni(111) by low temperature chemical vapor deposition. Applied Physics Letters.

[34] Xu X, Zhang Z, Dong J, et al. Ultrafast epitaxial growth of metre-sized single-crystal graphene. Science Bulletin. 2017;**62**(15):1074-1080

[35] Wang H, Wang G, Bao P, et al. Controllable synthesis of submillimeter single-crystal monolayer graphene domains on copper foils by suppressing nucleation. Journal of the American Chemical Society. 2012;**134**(8):3627

[36] Fan L, Li Z, Li X, et al. Controllable growth of shaped graphene domains by atmospheric pressure chemical vapour deposition. Nanoscale. 2011;**3**(12):4946-4950

[37] Liu L, Zhou H, Cheng R, et al. High-yield chemical vapor deposition growth of high-quality large-area AB-stacked bilayer graphene. Acs Nano. 2012;**6**(9):8241-8249

[38] Yu Q, Jauregui LA, Wu W, et al. Control and characterization of individual grains and grain boundaries in graphene grown by chemical vapour deposition. Nature Materials. 2010;**10**(6):443-449

[39] Luo B, Chen B, Meng L, Geng D, et al. Layer-stacking growth and electrical transport of hierarchical graphene architectures. Advanced Materials. 2014;**26**:3218-3224

[40] Wu B, Geng D, Xu Z et al. Self-organized graphene crystal patterns. NPG Asia Materials. 2013;**S**:e36

[41] Lotya M, Hernandez Y, King PJ, et al. Liquid phase production of graphene by exfoliation of graphite in surfactant/water solutions. Journal of the American Chemical Society. 2009;**131**(10):3611-3620

[42] Hummers WS, Offeman RE. Preparation of graphitic oxide. Journal of the American Chemical Society. 1958;**80**(6):1339

[43] Song HS, Li SL, Miyazaki H, Sato S, Hayashi K, Yamada A, et al. Origin of the relatively low transport mobility of graphene grown through chemical vapor deposition. Scientific Reports. UK. 2012;**2**:337

[44] Muñoz R, Gómez-Aleixandre C. Review of CVD synthesis of graphene. Chemical Vapor Deposition. 2013;**19**(10-11-12):297-322

[45] Li X, Cai W, Colombo L, Ruoff RS. Evolution of graphene growth on Ni and Cu by carbon isotope labeling. Nano Letters. 2009;**9**:4268-4272

[46] Van Bommel AJ, Crombeen JE, Van Tooren A. LEED and Auger electron observations of the SiC(0001) surface. Surface Science. 1975;**48**(2):463-472

[47] Berger C et al. Ultrathin epitaxial graphite: 2D electron gas properties and a route toward graphene-based nanoelectronics. The Journal of Physical Chemistry B. 2004;**108**(52): 19912-19916

[48] Zhang X, Li H, Ding F. Self-assembly of carbon atomson transition metal surfaces—Chemical vapor deposition growth mechanism of graphene. Advanced Materials. 2014;**26**:5488-5495

[49] Kwon S-Y, Ciobanu CV, Petrova V, Shenoy VB, Bareno J, Gambin V, et al. Growth of semiconducting graphene on palladium. Nano Letters. 2009;**9**:3985-3390

[50] Losurdo M, Giangregorio MM, Capezzuto P, Bruno G. Graphene CVD growth on copper and nickel: Role of hydrogen in kinetics and structure. Physical Chemistry Chemical Physics. 2011;**13**:20836-20843

[51] Nie S, Bartelt NC, Wofford JM, Dubon OD, McCarty KF, Thurmer K. Scanning tunneling microscopy study of graphene on Au(111): Growth mechanisms and substrate interactions. Physical Review B. 2012;**85**:205406

[52] Li Z et al. Low-temperature growth of graphene by chemical vapor deposition using solid and liquid carbon sources. ACS Nano. 2011;**5**(4):3385-3390

[53] Jang J et al. Low-temperature-growth continuous graphene films from benzene by chemical vapor deposition at ambient pressure. Scientific Reports. 2015;**5**:17955

[54] Jiang B-B, Pan M, Wang C, Wu MH, Vinodgopal K, Dai G-P. Controllable synthesis of circular graphene domains by atmosphere pressure chemical vapor deposition. Journal of Physical Chemistry C. 2018;**122**:13572-13578

[55] Weatherup RS, Dlubak B, Hofmann S. Kinetic control of catalytic CVD for high-quality graphene at low temperatures. ACS Nano. 2012;**6**(11):9996-10003

[56] Yokoyama H, Numakura H, Koiwa M. The solubility and diffusion of carbon in palladium. Acta Materials. 1998;**46**(8):2823-2830

[57] Sutter PW, Flege JI, Sutter EA. Epitaxial graphene on ruthenium. Nature Materials. 2008;**7**(5):406-411

[58] Shu N, Bartelt NC, Starodub E, Walter A, Bostwick A, Rotenberg E. Growth from below: Graphene multilayers on Ir(111). United States. 2011. DOI: 10.1021/nn103582g

[59] Loginova E, Bartelt NC, Feibelman PJ, McCarty KF. Evidence for graphene growth by C cluster attachment. New Journal of Physics. 2009;**11**:063046 (093026)

[60] Yan Z, Lin J, Peng ZW, Sun ZZ, Zhu Y, Li L, Xiang CS, Samuel EL, Kittrell C, Tour JM. ACS Nano. 2012;**10**(6):126-132

[61] Zhang Y, Li Z, Kim P, Zhang LY, Zhou CW. ACS Nano. 2012;**7**(6):871-877

[62] Meca E, Lowengrub J, Kim H, Mattevi C, Shenoy VB. Epitaxial graphene growth and shape dynamics on copper: Phase-field modeling and experiments. Nano Letters. 2013; **11**(13):5692-5697

[63] Jacobberger RM, Arnold MS. Graphene growth dynamics on epitaxial copper thin films. Chemistry of Materials. 2013;**6**(25):3040-3047

[64] Park HJ et al. Growth and properties of few-layer graphene prepared by chemical vapor deposition. Carbon. 2010;**48**(4):1088

[65] Zhang X, Wang L, Xin JH, et al. The role of hydrogen in graphene chemical vapor deposition (CVD) growth on copper surface. Journal of the American Chemical Society. 2014;**136**(8):3040-3047

[66] Weiss SJ, Young J, Lobuglio AF, et al. Role of hydrogen peroxide in neutrophil-mediated destruction of cultured endothelial cells. Journal of Clinical Investigation. 1981;**68**(3):714-721

[67] Liu Q, Gong Y, Wilt JS, et al. Synchronous growth of AB-stacked bilayer graphene on Cu by simply controlling hydrogen pressure in CVD process. Carbon. 2015;**93**:199-206

[68] Shu H, Tao XM, Ding F. What are the active carbon species during graphene chemical vapor deposition growth? Nanoscale. 2015;**7**(5):1627-1634

[69] Li K, He C, Jiao M, et al. A first-principles study on the role of hydrogen in early stage of graphene growth during the CH4 dissociation on Cu(111) and Ni(111) surfaces. Carbon. 2014;**74**(10):255-265

[70] Gan CK, Srolovitz DJ. First-principles study of graphene edge properties and flake shapes. 2010;**81**(12):760-762

[71] Geng D, Wu B, Guo Y, et al. From the cover: Uniform hexagonal graphene flakes and films grown on liquid copper surface. Proceedings of the National Academy of Sciences of the United States of America. 2012;**109**(21):7992-7996

[72] Luo B, Gao E, Geng D, et al. Etching-controlled growth of graphene by chemical vapor deposition. Chemistry of Materials. 2017;**29**(3):1022-1027

[73] Zhang H, Zhang Y, Zhang Y, et al. Realizing controllable graphene nucleation by regulating the competition of hydrogen and oxygen during chemical vapor deposition heating. Physical Chemistry Chemical Physics. 2016;**18**(34):23638-23642

[74] Ma T, Ren W, Liu Z, et al. Repeated growth-etching-regrowth for large-area defect-free single-crystal graphene by chemical vapor deposition. ACS Nano. 2014;**8**(12):12806-12813

[75] Da HJ, Kang C, Kim M, et al. Effects of hydrogen partial pressure in the annealing process on graphene growth. Journal of Physical Chemistry C. 2015;**118**(7):3574-3580

[76] Hu B, Jin Y, Guan D, et al. H2-dependent carbon dissolution and diffusion-out in graphene chemical vapor deposition growth. Journal of Physical Chemistry C. 2015;**119**(42): 585080264

[77] Jin Y, Hu B, Wei Z, et al. Roles of H2 in annealing and growth times of graphene CVD synthesis over copper foil. Journal of Materials Chemistry A. 2014;**2**(38):16208-16216

[78] Wu B, Geng D, Guo Y, et al. Equiangular hexagon-shape-controlled synthesis of graphene on copper surface. Advanced Materials. 2011;**23**(31):3522-3525

[79] Zhao P, Cheng Y, Zhao D, et al. The role of hydrogen in oxygen-assisted chemical vapor deposition growth of millimeter-sized graphene single crystals. Nanoscale. 2016;**8**(14): 7646

[80] Shin YC, Kong J. Carbon. 2013;**59**:439

[81] Zhang XY, Wang L, Xin J, Yakobson BI, Ding F. Journal of American Chemical Society. 2014;**18**:136

[82] Hao Y, Ruoff RS. The role of surface oxygen in the growth of large single-crystal graphene on copper. Science. 2013;**342**(6159):720-723

[83] Liang T, Luan C, Chen H, et al. Exploring oxygen in graphene chemical vapor deposition synthesis. Nanoscale. 2017;**9**(11):3719

[84] Zhang Y, Li Z, Kim P, Zhang LY, Zhou CW. ACS Nano. 2012;**7**:6

[85] Wang ZG, Chen YF, Li PJ, et al. Effects of methane flux on structural and transport properties of CVD-grown graphene films. Vacuum. 2012;**86**(7):895

[86] Nguyen VL, Shin BG, Duong DL, et al. Seamless stitching of graphene domains on polished copper (111) foil. Advanced Materials. 2015;**27**(8):1376-1382

[87] Yamada T, Ishihara M, Hasegawa M. Large area coating of graphene at low temperature using a roll-to-roll microwave plasma chemical vapor deposition. Thin Solid Films. 2013;**532**(532):89-93

[88] Juang ZY, Wu CY, Lu AY, et al. Graphene synthesis by chemical vapor deposition and transfer by a roll-to-roll process. Carbon. 2010;**48**(11):3169-3174

[89] Bae S, Kim H, Lee Y, Xu X, Park J-S, Zheng Y, et al. Roll-to-roll production of 30-inch graphene films for transparent electrodes. Nature Nanotechnology. 2010;**5**:574

[90] Kraus J, Böbel M, Günther S. Suppressing graphene nucleation during CVD on polycrystalline Cu bycontrolling the carbon content of the support foils. Carbon. 2016;**96**:153-165

[91] Zhang H, Zhang Y, Wang B, Chen Z, Sui Y, Zhang Y, et al. Journal of Electronic Materials. 2015;**44**:79-86

[92] Kobayashi T, Bando M, Kimura N, Shimizu K, Kadono K, Umezu N, et al. Production of a 100-m-long high-quality graphene transparent conductive film by roll-to-roll chemical vapor deposition and transfer process. Applied Physics Letters. 2013;**102**:023112

[93] Choi W, Lahiri I, Seelaboyina R, Kang YS. Synthesis of graphene and its applications: A review. Critical Reviews in Solid State and Materials Sciences. 2010;**35**:52-71

[94] Li X, Sun X, et al. A simple one-pot strategy for synthesizing ultrafine SnS2 nanoparticle/graphene composites as anodes for lithium/sodium-ion batteries. ChemSuschem. 2018;**11**(9):1390-1575

[95] Molina-Valdovinos S et al. Low-dimensional thermoelectricity in graphene: The case of gated graphene superlattices. Physica E: Low-dimensional Systems and Nanostructures. 2018;**101**:188-196

[96] Arapov K et al. Conductive screen printing inks by gelation of graphene dispersions. Advanced Functional Materials. 2016;**26**(4):586-593

[97] Wang JJ, Zhu MY, Outlaw RA, Zhao X, Manos DM. Free-standing sub-nanometer graphite sheets. Applied Physics Letters. 2004;**85**:1265

Chapter 6

Atmospheric Pressure Chemical Vapor Deposition of Graphene

Phuong V. Pham

Additional information is available at the end of the chapter

http://dx.doi.org/10.5772/intechopen.81293

Abstract

Recently, graphene has gained significant interest owing to its outstanding conductivity, mechanical strength, thermal stability, etc. Among various graphene synthesis methods, atmospheric pressure chemical vapor deposition (APCVD) is one of the best syntheses due to very low diffusivity coefficient and a critical step for graphene-based device fabrication. High-temperature APCVD processes for thin film productions are being recognized in many diversity technologies such as solid state electronic devices, in particular, high quality epitaxial semiconductor films for silicon bipolar and metal oxide semiconductor (MOS) transistors. Graphene-based devices exhibit high potential for applications in flexible electronics, optoelectronics, and energy harvesting. In this chapter, recent advances of APCVD-based graphene synthesis and their related applications will be addressed.

Keywords: graphene, atmospheric pressure chemical vapor deposition (APCVD), single-layer graphene (SLG), bilayer graphene (BLG), atmosphere pressure, large-scale

1. Introduction

Single-layer graphene (SLG), bilayer graphene (BLG), and multi-layer graphene (MLG) films have been regarded as optimal materials for electronics and optoelectronics owing to their excellent electrical properties and their ability to integrate with current top-down device fabrication technology [1–95]. Since the beginning of the twenty-first century, the interest in graphene materials has drastically increased, which is apparent in the number of annual publications on graphene (**Figure 1**). Until now, various strategies including chemical vapor deposition (CVD) [28], liquid and mechanical exfoliation from graphite [23, 29, 30], epitaxial growth on crystal substrate [31–34], or solution-based processes on graphene oxides (GOs)

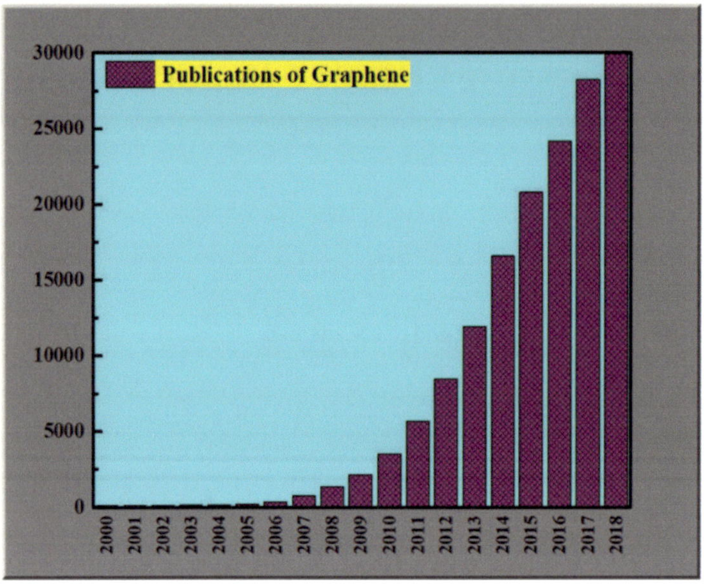

Figure 1. Graphene publications from 2000 to 2018. Source: ISI Web of Science (search: Topic = Graphene).

[35–41]. They have been investigated for obtaining graphene layers. In particular, recent advances in CVD growth have successfully led to large-scale graphene production on metal substrates [1, 27, 42–48], driven by the high demand for utilizing graphene in possible applications of current complementary metal-oxide-semiconductor (CMOS) technology such as radio-frequency transistors, optical devices, and deposition processes [2].

High-quality large-scale graphene has been synthesized on conducting metallic substrates by using the catalytic CVD growth approach, which promoted a wide range of graphene-based device applications [1, 28, 42–49]. However, graphene grown on a metal substrate needs to be transferred onto dielectric substrates for electronic applications. Although various approaches, such as wet etching/transfer [45], mechanical exfoliation/transfer [23, 29, 30], bubbling transfer [50], electrochemical delamination [51–53], for transferring from the catalytic metal growth substrates to dielectric device substrates have been developed; none of these approaches is free from degradation of the transferred graphene. For example, 'wet etching and transfer,' the most widely used transfer approach, is a serial process, which includes encapsulation of the graphene surface with polymer support.

Growing through CVD is for the production of films or for coating of metal, semiconductor, crystalline, and vitreous formed-compounds in either, occupying high-purity as well as desirable characteristics. In addition, the creation of controllable film with varying stoichiometry results CVD uniquely among deposition approaches. Other advantages include reasonably low cost of the equipment and operating expenses, suitability for semicontinuous operation. Consequently, the variants of CVD have been developed recently such as low-pressure chemical vapor deposition (LPCVD), APCVD, plasma-enhanced chemical vapor deposition (PECVD), and laser-enhanced chemical vapor deposition (LECVD). The hybrid system represents for both CVD and physical vapor deposition (PVD) have also discovered.

Among the utilized synthesis methods, APCVD has been considered as the most potential and medium cost one for large-scale high-quality graphene on various metallic substrates including Pt [54], Ir [42], Ni [7], and Cu [45]. Especially, Cu is considered as the best choice owing to low-carbon solubility, well-controlled surface, and inexpensive for growing monolayer graphene [45, 55]. Many efforts have made for obtaining large-scale single crystal graphene with as less grain boundaries as possible. There are two general methods to realize: the first method associates with the growth of single domains with possible enlargement [54, 56–63]. Despite the centimeter-scale domains have achieved, this method is not the bright candidate in practice for large-scale growth because of the difficult quantity control of nucleation seeds as well as unclear self-limiting growth factors resulting worse and requiring very long growth time (over 24 h) [64, 65]. The second method is the alignment of graphene domain orientations on arbitrary substrates and then to atomically stitch them to form uniform single crystalline graphene [5, 66]. This seems to be ideal for growing large scale.

The CVD method creates large-scale graphene but polycrystalline morphology, which includes different oriented domains. Such the orientated disorders inevitably lead to graphene grain boundaries (GGBs) formation at intercalated interface of domain [67–69]. GGBs include a series of nonhexagonal rings of pentagons, heptagons, and octagons. Intrinsic graphene has high conductivity and chemical inert, however, the appearance of atomically defect lines and GGBs in graphene could remarkably modulate its features such as mechanics [69, 70], electrics [71, 72], chemistry [73, 74], and magnets [75]. These defect lines have extraordinary features and rely on the atomical configuration at GGBs and the crystallinity of mono-grains; moreover, they could be modulated through different function groups [75]. Therefore, the investigation on the orientations and boundaries of the grains is the key to comprehend the underlying features as well as to realize compatible applications of graphene.

Conventionally, graphene is grown on metal foils (Cu, Ni, Pt, etc.) through APCVD, LPCVD, or PECVD. The temperatures, pressures, and concentrations of precursor gases inside the furnace play key roles on the graphene growth and quality. These factors need be optimized to get the desirable growth results. In general, surface morphology is studied via optical microscope (OM) and scanning electron microscopy (SEM); the graphene quality is examined through Raman spectra, UV-visible spectroscopy, and transmission electron microscopy (TEM); and sheet resistance (R_s) of synthesized graphene is measured utilizing four-point probe technique. Almost graphene films synthesized via APCVD are monolayer with better quality in atmospheric condition compared with LPCVD or PECVD. The synthesis study using APCVD would be greatly significant in the desirable growth of graphene and other related materials. In this chapter, recent advances of APCVD-based graphene synthesis with practical concerns about chemical vapor transport, deposition process as well as their device applications will be addressed.

2. Growth mechanism of APCVD-based graphene

APCVD growth of graphene is a chemical synthesis process at atmosphere pressure for the formation of SLG or FLG on an arbitrary substrate by exposing the substrate to the gas-phase precursors at controlled reaction conditions [76]. Owing to the versatile nature of APCVD, intricately mixed homogeneous gas-phase and heterogeneous surface reactions are involved [77]. In general, as the partial pressure and/or temperature in the reaction substances are increased, homogeneous gas-phase reactions and the resulting homogeneous nucleation became significant [77]. To grow a high-quality graphene layer, this homogeneous nucleation needs to be minimized [77]. A general mechanism for APCVD-based graphene growth on catalytic metal substrates, for the growth of uniform and highly crystalline graphene layer on the surface, includes eight steps as follow: (1) mass transport of the reactant, (2) reaction of the film precursor, (3) diffusion of gas molecules, (4) adsorption of the precursor, (5) diffusion of the precursor into substrate, (6) surface reaction, (7) desorption of the product, and (8) removal of the by-product (**Figure 2**) [78, 96].

It has been a general observation in most of the experimental works that the LPCVD produces nonuniform thick graphene layers. However, as high-quality single-layer graphene growth, APCVD would be a better choice due to reduced mean free path/diffusion coefficient of the reactive species on the catalyst [79]. APCVD shows a much lower diffusivity coefficient: $D_g \sim 1/(\text{total pressure})$ compare with LPCVD. As the result, the small multilayer graphene islands is less appearing on a full large monolayer graphene island via APCVD growth process [79] compared with LPCVD.

In another graphene growth mechanism model, hydrocarbon molecules are absorbed as well as dissociated on Cu forming active carbon species by dehydrogenation reaction (**Figure 3**). These species are diffused on two sides of Cu foil and agglomerated on its active sites to form graphene nucleation seeds. Actually, introducing H_2 gas is mandatory as a key role in graphene growth for most CVD approaches. The overall processes of graphene growth on Cu are described (**Figure 3**) [89]. Generally, there are three main expected steps: (i) adsorp-decompose, (ii) diffuse-desorb, and (iii) nucleate-grow (**Figure 3**) [89]. The active carbon species are commonly not stable and easily agglomerate with thermodynamically stable species on active sites to form graphene nucleation seeds by the reactions: $(CH_x)_s$ + graphene ➔ (graphene-C) + $(CH_x)_s$ with x = 1, 2, 3 [89]. Till those

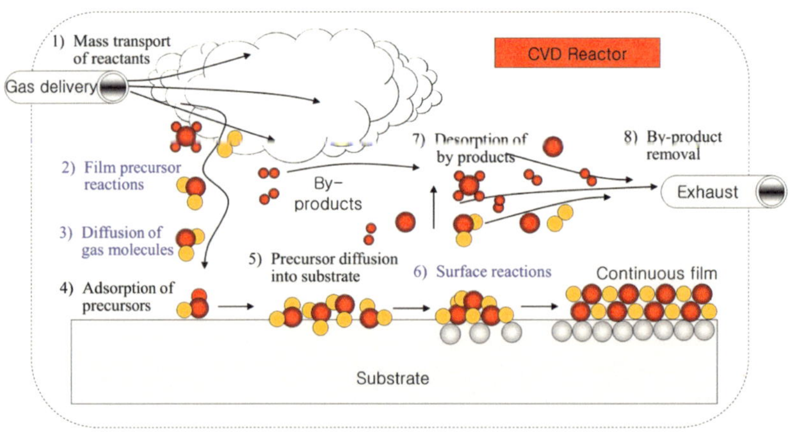

Figure 2. Diagram of CVD growth mechanism (APCVD and LPCVD) of graphene: transport and reaction processes. Reproduced with permission from [78, 96]. Copyright 2011, Freund Publishing.

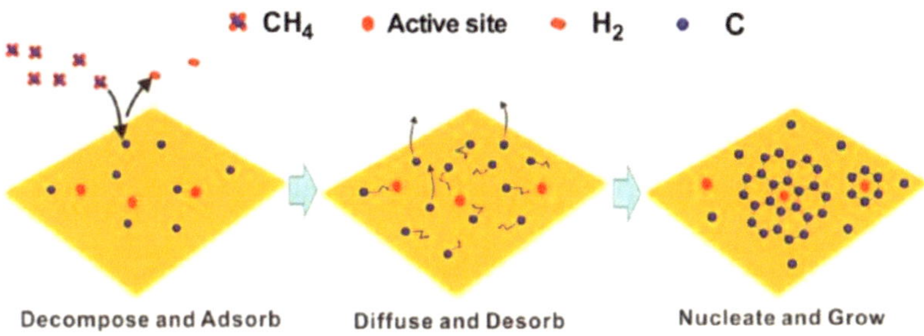

Figure 3. Schematic for graphene growth mechanism on Cu substrate. Reproduced with permission from [89]. Copyright 2012, American Chemical Society.

seeds formed, almost the active carbon species are incorporated and captured at surface/interface of graphene lattice. In addition, H_2 precursor plays more role as an etchant and controls the size as well as shape of graphene domains via reactions: H_s + graphene ➔ (graphene-C) + $(CH_x)_s$ with x = 1, 2, 3 [89].

Typically, APCVD growth of 2D materials (e.g. graphene) involves catalytic activation of chemical reactions of precursors at the growth substrate surface/interface in a properly designed environment. In general, the roles of precursors, conditions (e.g. fast growth rates, large domain size, or very high crystalline quality), atmosphere, substrates, and catalysts are the key factors affecting the final quality of the grown 2D materials. So far, significant efforts have been made to prepare high crystalline 2D materials (e.g. graphene), but many challenges are still ahead. For example, due to the rough feature of catalytic metal surface, growth of uniform and high-quality graphene is considerably difficult. The 2D material research community is also interested in new precursors (e.g. solid precursor only, gas precursor, or solid precursor mixed with certain solvents) that could form uniformly high-quality graphene with minimal defect density. Another question is the effect of growth rate on the catalytic metal surface on the quality of graphene. Currently, it is difficult to give an exact answer, as investigations are progressing at an exponential rate. To date, the understanding of the concept of the general mechanism of the APCVD growth of graphene is still not yet adequate, neither experimentally nor theoretically. Thus, understanding the graphene growth mechanism and the effect of various growth conditions will be of significant interest to the 2D material research community to obtain large scale, high-quality graphene.

3. APCVD growth of graphene

CVD is a thin solid film deposition process of vapor species through suitable chemical reaction. The deposition needs low-carbon solubility substrate in high-temperature region [80]. To date, CVD is the sole approach, which could produce high-quality graphene with ultra-large size [45, 81]. For the first experiment, researchers successfully synthesized graphene films via CVD on Ni and Cu catalytic substrates [44, 82, 83]. The significant progress for large-size and high-quality graphene films has also well done [81, 83].

APCVD requires high temperature (~1000°C) for graphene synthesis. It is experimentally expensive; therefore, it requires the CVD equipment more sophisticatedly. Unfortunately, this is an obstacle for direct-growth of graphene on insulating device substrate (e.g., SiO_2), for instance, it can produce unavoidable physical damages around 1000°C and dramatically degrades the quality of synthesized graphene. As a result, the deposition of graphene on insulating substrates at reduced temperatures becomes very necessary [84, 85]. The features of graphene will change corresponding with the number of layers. Hence, the electronic, optical, mechanical, and other features of graphene could be tuned by controlling the number of layers, or by adjusting the experimental conditions. Currently, it is still the challenge to exactly atomic-scale control the number of graphene layers [86–88]. The proposed synthesis method is composed by four stages (cleaning, precursor injection, reaction time, and cooling). The temperature and flow rate of carbon precursor were the studied parameters. In **Figure 4a, b**, the common CVD systems and differences are shown [90].

In 2016, Wang et al. obviously showed millimeter-scale graphene single crystals synthesized on Cu through APCVD (**Figure 5**) [91]. The possible mechanism will be based on graphene nucleation and kinetic growth. At the stage of nucleation, thermal decomposition of oxide layer leads to O_2 desorption at high temperature at front side of Cu and dominates the temperature dependence of nucleation density. The graphene island growth is edge-limited on two sides of Cu at various enlargement rates. The roughness of support substrates (quartz, sapphire) also affects the graphene deposition. After optimized annealing and polished

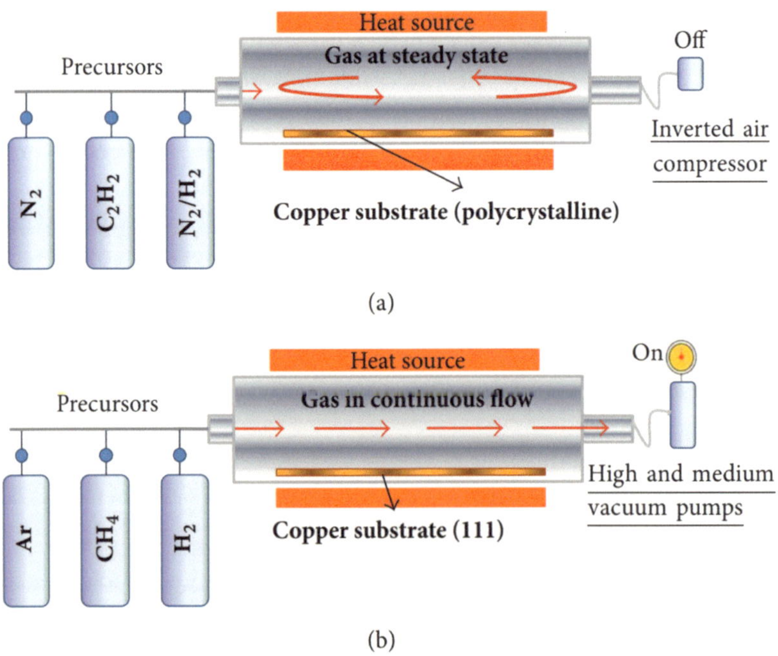

Figure 4. Schematic of graphene synthesis by (a) APCVD using polycrystalline Cu substrate and gas discontinuous flow and (b) high vacuum using single crystalline Cu(111) substrate and gas continuous flow. Reproduced with permission from [90]. Copyright 2018, Hindawi Publishing.

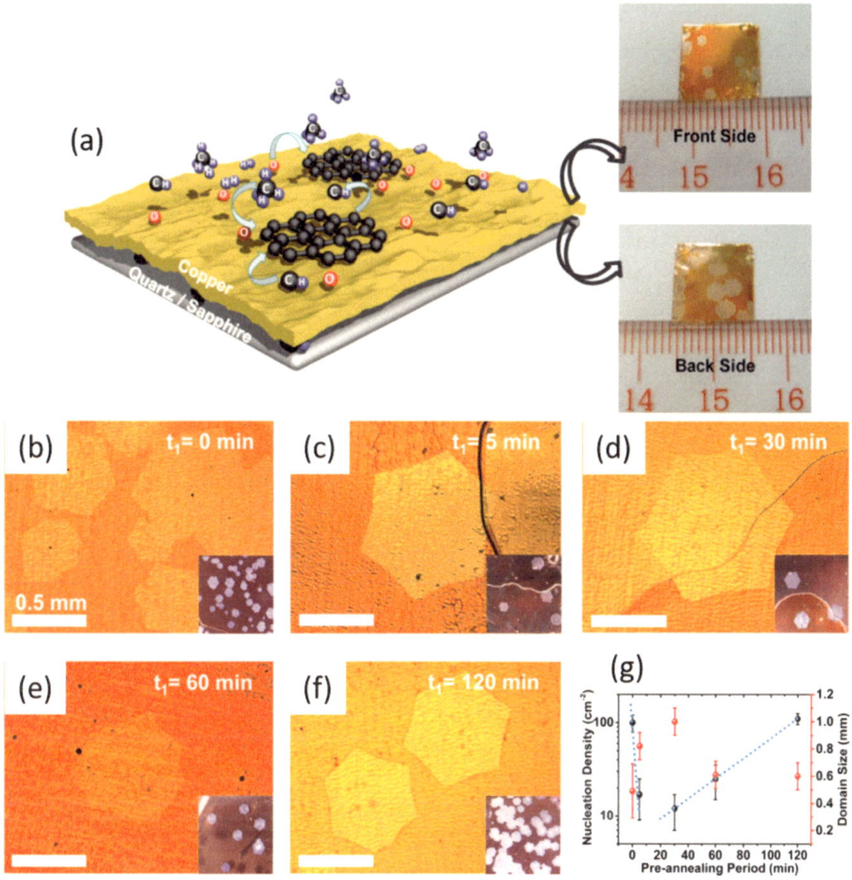

Figure 5. (a) Schematic for APCVD-grown graphene on Cu located on quartz or sapphire substrates. Insets in (a) are OM images of graphene/Cu at front side and back side. OM images of grown graphene on Cu foils with various pre-annealing times of (b) 0, (c) 5, (d) 30, (e) 60, (f) 120 min at the same growth time (60 min). (g) Evolution of nucleation density and size of graphene domains as a function of pre-annealing times. Reproduced with permission from [91]. Copyright 2016, American Chemical Society.

support substrate, the isolated graphene islands (~3 mm) were produced with a growth rate (25 μm/min). The domains were uniformly single-crystalline graphene with good mobility (~4900 cm^2/V.s) at room temperature. **Figure 5b-f** shows OM images of graphene domains/ Cu for 60 min growth at various Ar pre-annealing times. The growth and cooling parameters maintained the same. Here, the Ar heating induced the growth of individual graphene domains with relatively low nucleation density (102 cm^{-2}) owing to the catalyst passivation of oxide layer on top of Cu. Through annealing in 30 min (**Figure 5d**), the nucleation density reduced ~12 nuclei/cm^2, and a grain size of ~1 mm was obtained after 60 min growth (**Figure 5e**). But the extending of annealing (120 min) caused the increase of nucleation density with reduced domain size (~0.6 mm) (**Figure 5f**). The evolutions of nucleation densities and domain sizes via annealing are shown in **Figure 5g**. The nucleation density reveals a nonmonotonic dependence of Ar-passivated annealing. Since the surface becomes flatter after annealing, the O-species also gradually desorb at surface, which leaves pristine Cu as active nucleation sites following the growth steps.

Thermodynamics of CVD-based graphene synthesis using catalytic Cu at particular temperature is same irrespective at atmosphere pressure (AP), low pressure (LP), or ultrahigh vacuum (UHV) [79]. But, the kinetics will vary on different steps. The kinetics of cooling rate, the pressure of CVD furnace has major ramification on the graphene growth rate, large-scale thickness uniformity, and the defect density. **Figure 6A** illustrates a steady state flow of a mixture of CH_4, H_2, and Ar precursors on Cu surface at ~1000°C [79]. The boundary layer because of steady state gas flow is stagnant. Firstly, the carbon species (1) diffuse via the boundary layer to the surface, then (2) will be adsorbed on the surface, (3) decompose for formation of active carbon species, (4) diffuse on/into the catalyst surface and forming the graphene lattice, (5) inactive species (e.g. H_2) get desorbed, forming H_2, then (6) diffuse away through boundary layer and swept away by the bulk gas flow [79]. Processes, which occur on/close the surface, are highly affected by substrate temperature. Generally, there are two fluxes of active species: flux of active species via boundary layer and at metal substrate surface forming graphene lattice (**Figure 6B**) [79]. The equations of these fluxes are $F_{mass\ transport} = h_g(C_g - C_s)$ and $F_{surface\ reaction} = K_sC_s$ where, $F_{mass\text{-}transport}$ is the flux of active species through the boundary layer, $F_{surface\text{-}reaction}$ is the flux of consumed active species at surface, h_g is the mass transport coefficient, K_s is the surface reaction constant, C_g is the concentration of gas in bulk, and C_s is the concentration of active species at the surface [79]. At high temperatures, under APCVD parameters, mass transport via boundary layer is rate limiting ($K_s \gg h_g$), and under LP and UHV parameters, the surface reaction is the rate limiting step ($h_g \gg K_s$) [79].

A diagram of designed-APCVD for graphene synthesis is described in **Figure 7**. This splits one gas inlet in two paths by means of glass valves: one for heating, annealing, and cooling (P1); and the other for graphene synthesis (P2). The synthesis temperature is 980–990°C in tubular furnace; during heating, the Ar-H_2 (5% of H_2) flow is 0.2 l/min via P1. Then Cu is annealed in 20–30 min. Then, the flow passed via alcohol container (P2) to desired growth rate, and the varied time depended on precursor type. Keeping the flow rate, the flux is to P1 and the sample is annealed in 10 min. As the result, this annealing step improved the graphene quality

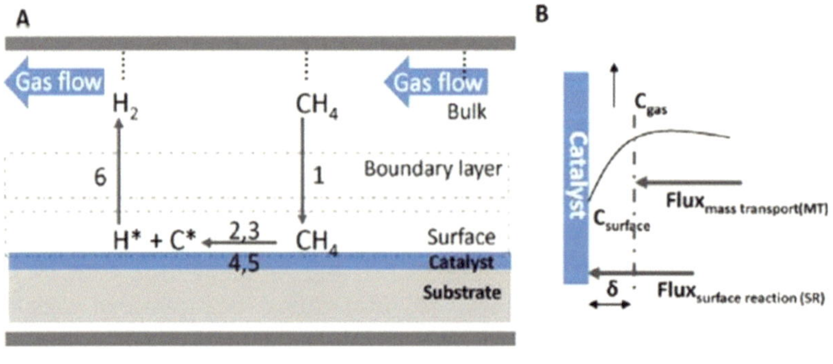

Figure 6. (A) Schematic for graphene growth mechanism using low carbon solid solubility catalysts (Cu) at atmosphere pressure (AP), low pressure (LP), or ultrahigh vacuum (UHV) environments. (B) Mass transport and surface reaction fluxes under steady state conditions. Reproduced with permission from [79]. Copyright 2010, American Chemical Society.

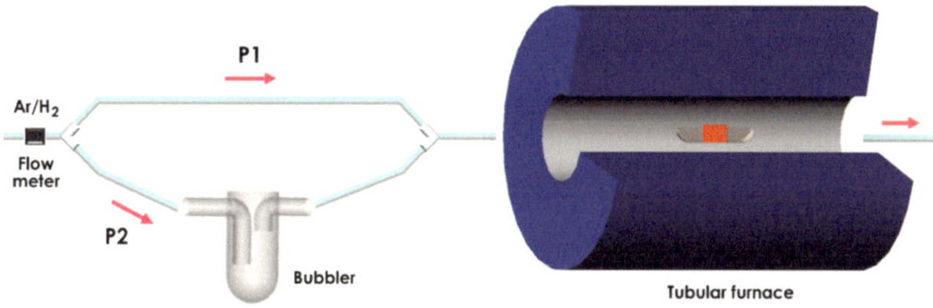

Figure 7. Graphene growth mechanism on new designed APCVD with a bubbler. A single gas inlet is split into two paths (P1 and P2) using glass. During heating/annealing/cooling, gas is passed through P1; while for graphene growth, gas flows through P2 and bubbler containing alcohol. Inside the tubular furnace a combustion boat is placed in the center containing a Cu foil. Reproduced with permission from [92]. Copyright 2013, Elsevier.

a lot. The temperature, the time, and flow rate permit a sophisticated controlling on carbon concentration toward tubular furnace (**Figure 7**) [92].

The additional oxide substrate (SiO_2, Al_2O_3) to continuously supply the oxygen (O_2) to Cu surface of APCVD graphene synthesis (**Figure 8a,b**) [93]. Cu was placed on oxide substrates away in 15 μm gap. Last report proved that the oxide shall slowly release O_2 at above 800°C) [97]. To clear it, Xu et al. carried out Auger electron spectroscopy (AES) element measurement on SiO_2, as the result, O_2 escaped from SiO_2 after annealing in UHV or CVD (**Figure 8c**) [93]. Despite the small amount of O_2 released, the O_2 concentration between the narrow gap (15 μm) of Cu and oxide substrate could be high because of the trapping effect; hence, the O_2 attachment to Cu surface significantly trapped. Beside AP condition, high CH_4 flow (5 sccm) and high CH_4/H_2 ratio (~1) ensure a sufficient supply of carbon for ultrafast domain synthesis. The large domains (~0.3 mm) appeared on the back surface of Cu (**Figure 8d**). Contrarily, graphene domains at front surface of Cu are 20 times smaller (~15 μm) (**Figure 8e**). All these domains are star-shaped. For further demonstration, when graphite is used as the supporting substrate, graphene domains on both sides of Cu are also star-shaped similar with the case of SiO_2 (**Figure 8g**) [93].

Figure 8h,i revealed OM images of graphene/Cu in 60 and 80 min under CH_4 0.000075 Pa [94]. The graphenes needed to place in air (200°C, 1 min) for chemical oxidation between Cu and air to emerge the nonoxidized-graphene domains for the observation by OM and human-eye. Graphene partly covered the Cu (~70%) after 60 min. Some isolated domains were ~1.5 mm (**Figure 8h**) [94]. But the color as well as contrast of sample were of no change after 80 min implying that Cu got full coverage by graphene via OM image (**Figure 8i**) [94]. The histogram in **Figure 8j** showed the growth distribution of domain sizes for 50–60 min [94]. The distribution of domain sizes increased via growth time; the domain sizes were 0.4–0.8 mm for 50–60 min. The time evolution of average domain size and coverage were clearly described in **Figure 8k** [94]. The domain sizes for 100% coverage after 80 min was ~1 mm [94].

A growth at APCVD in Mohsin et al. is to avoid Cu evaporation from the foil substrate, which commonly appears in LPCVD case [95, 98]. Regarding **Figure 9,** firstly Cu is placed on a tungsten (W) foil for preventing the dewetting of liquid Cu on quartz. Then, it was heated up 1100°C, which is higher melting point of Cu for 30 min under Ar (940 sccm) and H_2 (60 sccm).

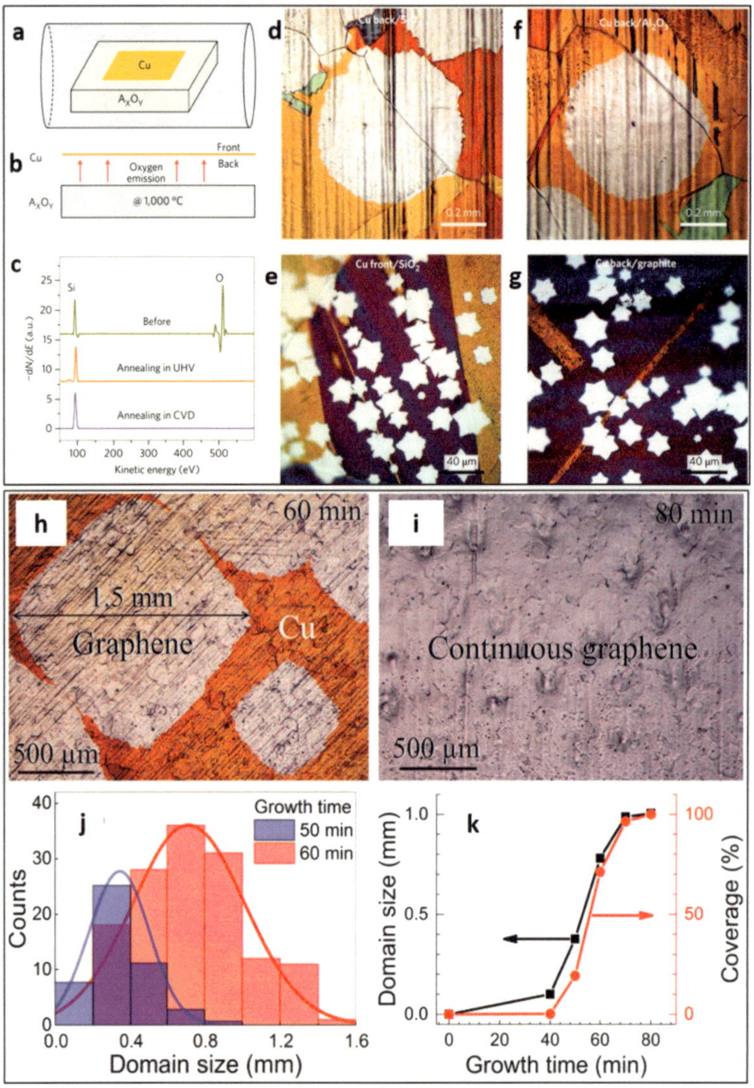

Figure 8. (a) Schematic of graphene growth on Cu using O_2-assisted APCVD. (b) Side view of (a). (c) Auger electron spectroscopy (AES) of SiO_2 before/after annealing at 1000°C in UHV and APCVD systems. The disappearance of O_2 peak after annealing proves the emission of O_2 from the oxide at high temperature. (d, e) OM images of graphene domains on the back (d) and front (e) Cu surfaces using fused silica as the supporting substrate. (f, g) OM images of graphene domains on the back Cu surfaces using sapphire (Al_2O_3) (f) or graphite (g). Reproduced with permission from [93]. Copyright 2016, Nature Publishing Group. (h, i) OM images of graphene/Cu after 60 and 80 min, respectively. APCVD conditions: 1 atm, 1030°C, 0.3 sccm CH_4, 80 sccm H_2, and 3920 sccm Ar. (j) Histogram of domain size for 50 and 60 min growth. (k) The graphene coverage and the average domain size as a function of growth time. Reproduced with permission from [94]. Copyright 2016, Nature Publishing Group.

Then, the temperature is lowered to 1075°C, and the Cu resolidified. Growth was carried out at this temperature, with 0.1% dilute CH_4 in Ar. On the right side of **Figure 9**, it shows OM image of two hexagon domains of graphene (>1 mm) which has similar morphology [95]. These domains obtained in Mohsin et al. are hexagons with very less roughness edges compared to previous reports [60, 89].

http://dx.doi.org/10.5772/intechopen.81293

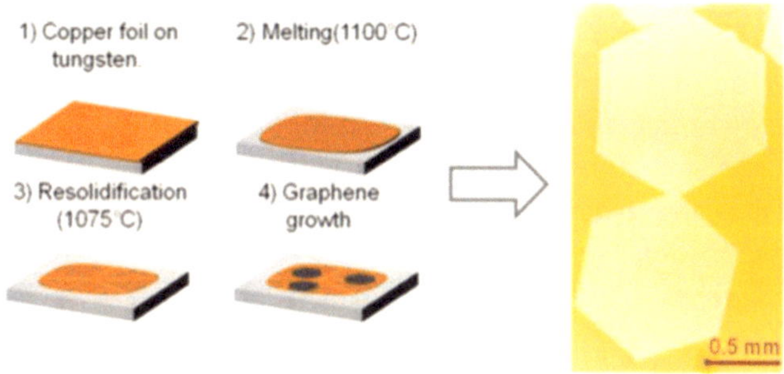

Figure 9. (a) Simple method to grow millimeter-size graphene single crystals on melted and resolidified Cu using APCVD. (b) OM image of the synthesized graphene domains. Reproduced with permission from [95]. Copyright 2013, American Chemical Society.

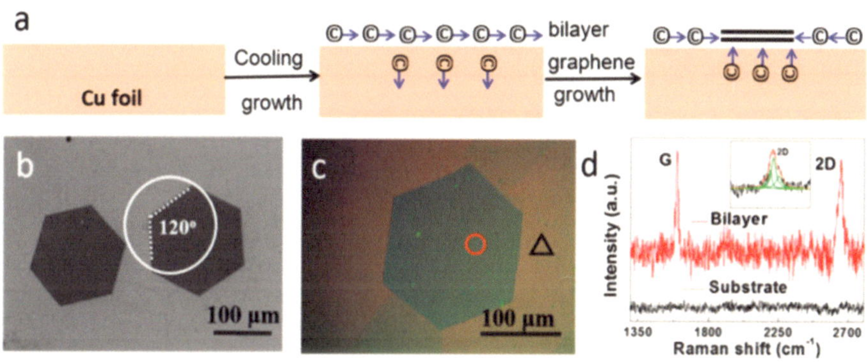

Figure 10. (a) Schematic of bilayer graphene growth using a cooling APCVD. (b, c) SEM and OM images of bilayer graphene domains, respectively. (d) Rama spectra of bilayer graphene. Reproduced with permission from [99]. Copyright 2016, American Chemical Society.

APCVD-based bilayer graphene growth is another great issue for optimum electronic and photonic devices because of higher carrier mobility and wider band gap by perpendicular electric field compared with single-layer graphene [99, 100]. The synthesis of bilayer graphene faced many drawbacks owing to limited grain size and nonsynchronic growth between the first and the second graphene layer. In 2016, Sun et al. reported a cooling-APCVD to growth of bilayer graphene on polycrystalline Cu (**Figure 10**) [99]. Here, the surface adsorption for decomposed carbons and phase segregation for dissolved carbon were shown. Consequently, the millimeter-scale hexagonal bilayer graphene (~1.0 mm) was produced. This study opens the possibility to grow centimeter-scale bilayer graphene for optimum graphene-based applications in recent future.

4. Conclusions

Strategies for direct graphene growth using APCVD method have been reviewed. The prospects of APCVD-grown graphene are bright and currently receiving considerable attention from the 2D material research community. Yet, understanding the growth process and conditions that affect the quality of graphene is not adequate.

Because APCVD growth depends on thermal decomposition of carbon resource, the growth rate is usually low and size of the graphene domain is small, resulting in growth of defective graphene layer. To date, the challenge remains for this research directions and large-scale high-quality graphene productions are still hard. In order to obtain more advanced results, an in-depth mechanism understanding of APCVD-based graphene growth is essential.

Cu is considered as the best substrate owing to low-carbon solubility, well-controlled surface, and inexpensive for growing monolayer graphene. Generally, graphene is synthesized on Cu via APCVD parameters utilizing CH_4 precursor revealed that the growth is different from single layer graphene at low CH_4 concentration to multi-layer domain on a single layer at higher CH_4 concentration. In particular, APCVD-based graphene growth at higher CH_4 concentration has no self-limiting compared with LPCVD, indicating that further investigation is needed to point out the growth mechanism. The kinetic processes utilizing low carbon solubility catalyst via APCVD parameters were presented.

Finally, APCVD-grown graphene on flexible/metal/dielectric substrates assisted by metal powder precursors (solid, gas, and solution) contained inside a sub-chamber for direct evaporation into an innovative APCVD main chamber allowing hexagonal domain formation on flexible/metal/dielectric substrates is the secret topic and needs to be exploited in the coming time.

Conflict of interest

There are no conflicts of interest to declare.

Author details

Phuong V. Pham[1,2]*

*Address all correspondence to: pvphuong@skku.edu

1 SKKU Advanced Institute of Nano Technology (SAINT), Sungkyunkwan University (SKKU), Suwon, Gyeonggi-do, South Korea

2 Center for Multidimensional Carbon Materials, Institute for Basic Science, Ulsan, South Korea

References

[1] Kim KS, Zhao Y, Jang H, Lee SY, Kim JM, Kim KS, et al. Large-scale pattern growth of graphene films for stretchable. Nature. 2009;**457**:706-710. DOI: 10.1038/nature07719

[2] Kim K, Choi JY, Kim T, Cho SH, Chung HJ. A role for graphene in silicon-based semiconductor devices. Nature. 2011;**479**:338-344. DOI: 10.1038/nature10680

[3] Duong DL, Han GH, Lee SM, Gunes F, Kim ES, Kim ST, et al. Probing graphene grain boundaries with optical miscroscopy. Nature. 2012;**490**:235-239. DOI: 10.1038/nature11562

[4] Choi J-Y. Graphene transfer: A stamp for all substrates. Nature Nanotechnology. 2013;**8**: 311-312. DOI: 10.1038/nnano.2013.74

[5] Lee JH, Lee EK, Joo WJ, Jang Y, Kim BS, Lim JY, et al. Wafer-scale growth of single-crystal monolayer graphene on reusable hydrogen-terminated germanium. Science. 2014;**344**: 286-289. DOI: 10.1126/science.1252268

[6] Shin HJ, Choi WM, Yoon SM, Han GH, Woo YS, Kim ES, et al. Transfer-free growth of few-layer graphene by self-assembled monolayers. Advanced Materials. 2011;**23**:4392-4397. DOI: 10.1002/adma.201102526

[7] Chae SJ, Güneş F, Kim KK, Kim ES, Han GH, Kim SM, et al. Synthesis of large-area graphene layers on poly-nickel substrate by chemical vapor deposition: Wrinkle formation. Advanced Materials. 2009;**21**:2328-2333. DOI: 10.1002/adma.200803016

[8] Güneş F, Shin HJ, Biswas C, Han GH, Kim ES, Chae SJ, et al. Layer-by-layer doping of few-layer graphene film. ACS Nano. 2010;**4**:4595-4600. DOI: 10.1021/nn1008808

[9] Pham VP, Jang HS, Whang D, Choi JY. Direct growth of graphene on rigid and flexible substrates: Progress, applications, and challenges. Chemical Society Reviews. 2017;**46**: 6276-6300. DOI: 10.1039/c7cs00224f

[10] Pham VP. A library of doped-graphene images via transmission electron microscopy. C. 2018;**4**:34. DOI: 10.3390/c4020034

[11] Pham VP, Nguyen MT, Park JW, Kwak SS, Nguyen DHT, Mun MK, et al. Chlorine-trapped CVD bilayer graphene for resistive pressure sensor with high detection limit and high sensitivity. 2D Materials. 2017;**4**:025049. DOI: 10.1088/2053-1583/aa6390

[12] Pham VP, Kim KN, Jeon MH, Kim KS, Yeom GY. Cyclic chlorine trap-doping for transparent, conductive, thermally stable and damage-free graphene. Nanoscale. 2014;**6**: 15301-15308. DOI: 10.1039/c4nr04387a

[13] Pham VP, Kim KH, Jeon MH, Lee SH, Kim KN, Yeom GY. Low damage pre-doping on CVD graphene/Cu using a chlorine inductively coupled plasma. Carbon. 2015;**95**: 664-671. DOI: 10.1016/j.carbon.2015.08.070

[14] Pham VP, Mishra A, Yeom GY. The enhancement of hall mobility and conductivity of CVD graphene through radical doping and vacuum annealing. RSC Advances. 2017;**7**: 16104-16108. DOI: 10.1039/c7ra01330b

[15] Pham VP. Cleaning of graphene surfaces by low pressure air plasma. Royal Society Open Science. 2018;**5**:172395. DOI: 10.1098/rsos.172395

[16] Pham VP, Kim DS, Kim KS, Park JW, Yang KC, Lee SH, et al. Low energy BCl3 plasma doping of few-layer graphene. Science of Advanced Materials. 2016;**8**:884-890. DOI: 10.1166/sam.2016.2549

[17] Kim KN, Pham VP, Yeom GY. Chlorine radical doping of a few layer graphene with low damage. ECS Journal of Solid State Science and Technology. 2015;**4**:N5095-N5097. DOI: 10.1149/2.0141506jss

[18] Pham VP. Chemical vapor deposited graphene synthesis with same-oriented hexagonal domains. Engineering Press. 2018;**1**:39-42. DOI: 10.28964/EngPress-1-107

[19] Pham VP. How can the nanomaterial surfaces be highly cleaned? Edelweiss Applied Science and Technology. 2018;**2**:184-186

[20] Pham VP. Layer-by-layer thinning of 2D materials. Edelweiss Applied Science and Technology. 2018;**2**:36-37

[21] Pham VP. Plasma-related graphene etching: A mini review. Journal of Materials Science and Engineering with Advanced Technology. 2018;**17**:91-106. DOI: 10.18642/jmseat_7100121943

[22] Pham VP. Graphene Etching: How Could it be Etched? Current Graphene Science; 2018 (Accepted)

[23] Geim AK, Novoselov KS. The rise of graphene. Nature Materials. 2007;**6**:183-191. DOI: 10.1038/nmat1849

[24] Neto AHC, Guinea F, Peres NMR, Novoselov KS, Geim AK. The electronic properties of graphene. Reviews of Modern Physics. 2009;**81**:109-162. DOI: 10.1103/RevModPhys.81.109

[25] Schwierz F. Graphene transistor. Nature Nanotechnology. 2010;**5**:487-496. DOI: 10.1038/nnano.2010.89

[26] Wilson NR, Macpherson JV. Carbon nanotube tips for atomic force microscopy. Nature Nanotechnology. 2009;**4**:483-491. DOI: 10.1038/nnano.2009.154

[27] Lin YM, Dimitrakopoulos C, Jenkins KA, Farmer DB, Chiu HY, Grill A, et al. 100GHz transistor from wafer-scale epitaxial graphene. Science. 2010;**327**:662. DOI: 10.1126/science.1184289

[28] Yan Z, Peng Z, Tour JM. Chemical vapor deposition of graphene single crystals. Accounts of Chemical Research. 2014;**47**:1327-1337. DOI: 10.1021/ar4003043

[29] Su CY, Lu AY, Xu Y, Chen FR, Khlobystov AN, Li LJ. High-quality thin graphene films from fast electrochemical exfoliation. ACS Nano. 2011;**5**:2332-2339. DOI: 10.1021/nn200025p

[30] Hernandez Y, Nicolosi V, Lotya M, Blighe FM, Sun Z, De S, et al. High-yield production of graphene by liquid-phase exfoliation of graphite. Nature Nanotechnology. 2008;**3**:563-568. DOI: 10.1038/nnano.2008.215

[31] Sutter PW, Flege JI, Sutter EA. Epitaxial graphene on ruthenium. Nature Materials. 2008; **7**:406-411. DOI: 10.1038/nmat2166

[32] Berger C, Song Z, Li X, Wu X, Brown N, Naud C, et al. Electronic confinement and coherence in patterned epitaxial graphene. Science. 2006;**312**:1191-1196. DOI: 10.1126/science.1125925

[33] Emtsev KV, Speck F, Seyller T, Ley L, Riley JD. Interaction, growth, and ordering of epitaxial graphene on SiC{0001} surfaces: A comparative photoelectron spectroscopy study. Physical Review B. 2008;**77**:155303. DOI: 10.1103/PhysRevB.77.155303

[34] Hass J, WAd H, Conrad EH. The growth and morphology of epitaxial multilayer graphene. Journal of Physics. Condensed Matter. 2008;**20**:323202. DOI: 10.1088/0953-8984/20/32/323202

[35] Su CY, Xu Y, Zhang W, Zhao J, Liu A, Tang X, et al. Highly efficient restoration of graphitic structure in graphene oxide using alcohol vapors. ACS Nano. 2010;**4**:5285-5292. DOI: 10.1021/nn10169m

[36] Williams G, Seger B, Kamat PV. TiO_2-graphene nanocomposites UV-assisted photocatalytic reduction of graphene oxide. ACS Nano. 2008;**2**:1487-1491. DOI: 10.1021/nn800251f

[37] Green AA, Hersam MC. Solution phase production of graphene with controlled thickness via density differentiation. Nano Letters. 2009;**9**:4031-4036. DOI: 10.1021/nl902200b

[38] Cote LJ, Kim F, Huang J. Langmuir-Blodgett assembly of graphite oxide singl layers. Journal of the American Chemical Society. 2008;**131**:1043-1049. DOI: 10.1021/ja806262m

[39] Li D, Müller MB, Gilje S, Kaner RB, Wallace GG. Processable aquaous dispersions of graphene nanosheets. Nature Nanotechnology. 2008;**3**:101-105. DOI: 10.1038/nnano.2007.451

[40] Gao W, Alemany LB, Ci L, Ajayan PM. New insights into the structure and reduction of graphite oxide. Nature Chemistry. 2009;**1**:403-408. DOI: 10.1038/nchem.281

[41] Joshi RK, Alwarappan S, Yoshimura M, Sahajwalla V. Graphene oxide: The new membrane material. Applied Materials Today. 2015;**1**:1-12. DOI: 10.1016/j.apmt.2015.06.002

[42] Coraux J, Diaye ATN, Busse C, Michely T. Structural coherency of graphene on Ir(111). Nano Letters. 2008;**8**:565-570. DOI: 10.1021/nl0728874

[43] Lee Y, Bae S, Jang H, Jang S, Zhu SE, Sim SH, et al. Wafer-scale synthesis and transfer of graphene films. Nano Letters. 2010;**10**:490-493. DOI: 10.1021/nl903272n

[44] Reina A, Jia X, Ho J, Nezich D, Son H, Bulovic V, et al. Large area, few-layer graphene films on arbitrary substrates by chemical vapor deposition. Nano Letters. 2008;**9**:30-35. DOI: 10.1021/nl801827v

[45] Li X, Cai W, An J, Kim S, Nah J, Yang D, et al. Large-area synthesis of high-quality and uniform graphene films on copper foils. Science. 2009;**324**:1312-1314. DOI: 10.1126/science.1171245

[46] Lee S, Lee K, Zhong Z. Wafer scale homogeneous bilayer graphene films by chemical vapor deposition. Nano Letters. 2010;**10**:4702-4707. DOI: 10.1021/nl1029978

[47] Yan K, Peng H, Zhou Y, Li H, Liu Z. Formation of bilayer Bernal graphene: Layer-by-layer epitaxy via chemical vapor deposition. Nano Letters. 2011;**11**:1106-1110. DOI: 10.1021/nl104000b

[48] Sun Z, Yan Z, Yao J, Beitler E, Zhu Y, Tour JM. Growth of graphene from solid carbon sources. Nature. 2010;**468**:549-552. DOI: 10.1038/nature09579

[49] Chen Y, Gong XZ, Gai JG. Progress and challenges in transfer of large-area graphene films. Advancement of Science. 2016;**3**:1500343. DOI: 10.1002/advs.201500343

[50] Sun J, Deng S, Guo W, Zhan Z, Deng J, Xu C, et al. Electrochemical bubbling transfer of graphene using a polymer support with encapsulated air gap as permeation stopping layer. Journal of Nanomaterials. 2016;**2016**:1-7. DOI: 10.1155/2016/7024246

[51] Cherian CT, Giustiniano F, Martin-Fernandez I, Andersen H, Balakrishnan J, Özyilmaz B. Bubble-free electrochemical delamination of CVD graphene films. Small. 2015;**11**: 189-194. DOI: 10.1002/smll.201402024

[52] Mafra D, Ming T, Kong J. Facial graphene transfer directly to target substrates with a reusable metal catalyst. Nanoscale. 2015;**7**:14807-14812. DOI: 10.1039/C5NR3892H

[53] Wang Y, Zheng Y, Xu X, Dubuisson E, Bao Q, Lu J, et al. Electrochemical delamination of CVD-grown graphene film: Toward the recyclable use of copper catalyst. ACS Nano. 2011;**5**:9927-9933. DOI: 10.1021/nn203700w

[54] Gao L, Ren W, Xu H, Jin L, Wang Z, Ma T, et al. Repeated growth and bubbling transfer of graphene with millimetre-size single-crystal grains using platinum. Nature Communications. 2012;**3**:699. DOI: 10.1038/ncomms1702

[55] Li X, Cai W, Colombo L, Ruoff RS. Evolution of graphene growth on Ni and Cu by carbon isotope labeling. Nano Letters. 2009;**9**:4268-4272. DOI: 10.1021/nl902515k

[56] Wang H, Wang G, Bao P, Yang S, Zhu W, Xie X, et al. Controllable synthesis of submilimeter single-crystal monolayer graphene domains on copper foils by suppressing nucleation. Journal of the American Chemical Society. 2012;**134**:3627-3630. DOI: 10.1021/ja2105976

[57] Eres G, Regmi M, Rouleau CM, Chen J, Ivanov IN, Puretzky AA, et al. Cooperative island growth of large-area single-crystal graphene on copper using chemical vapor deposition. ACS Nano. 2014;**8**:5657-5669. DOI: 10.1021/nn500209d

[58] Wang C, Chen W, Han C, Wang G, Tang B, Tang C, et al. Growth of millimeter-size single crystal graphene on Cu foils by circumfluence chemical vapor deposition. Scientific Reports. 2014;**4**:4537. DOI: 10.1038/srep04537

[59] Li X, Magnuson CW, Venugopal A, Tromp RM, Hannon JB, Vogel EM, et al. Large-area graphene single crystals grown by low-pressure chemical vapor deposition of methane on copper. Journal of the American Chemical Society. 2011;**133**:2816-2819. DOI: 10.1021/ja109793s

[60] Chen S, Ji H, Chou H, Li Q, Li H, Suk JW, et al. Milimeter-size single-crystal graphene by suppressing evaporative loss of Cu during low pressure chemical vapor deposition. Advanced Materials. 2013;**25**:2062-2065. DOI: 10.1002/adma.201204000

[61] Mohsin A, Liu L, Liu P, Deng W, Ivanov IN, Li G, et al. Synthesis of miloimeter-size hexagon-shaped graphene single crystals on resolidified copper. ACS Nano. 2013;**7**:8924-8931. DOI: 10.1021/nn4034019

[62] Wu T, Ding G, Shen H, Wang H, Sun L, Jiang D, et al. Triggering the continuous growth of growth toward milimeter-sized grains. Advanced Functional Materials. 2013;**23**: 198-203. DOI: 10.1002/adfm.201201577

[63] Gan L, Luo Z. Turning off hydrogen to realize seeded growth of subcentimeter single-crystal graphene grains on copper. ACS Nano. 2013;**7**:9480-9488. DOI: 10.1021/nn404393b

[64] Zhou H, Yu WJ, Liu L, Cheng R, Chen Y, Huang X, et al. Chemical vapour deposition growth of large single crystals of monolayer and bilayer graphene. Nature Communications. 2013;**4**:2096. DOI: 10.1038/ncomms3096

[65] Hao Y, Bharathi MS, Wang L, Liu Y, Chen H, Nie S, et al. The role of surface oxygen in the growth of large single-crystal gtaphene on copper. Science. 2013;**342**:720-723. DOI: 10.1126/science.1243879

[66] Nguyen VL, Shin BG, Duong DL, Kim ST, Perello D, Lim YJ, et al. Seamless stiching of graphene domains on polished copper (111) foil. Advanced Materials. 2015;**27**:1376-1382. DOI: 10.1002/adma.201404541

[67] Kim K, Lee Z, Regan W, Kisielowski C, Crommie MF, Zettl A. Grain bourdary mapping in polycrystalline graphene. ACS Nano. 2011;**5**:2142-2146

[68] Huang PY, Vargas CSR, van der Zande AM, Whitney WS, Levendorf MP, Kevek JW, et al. Grains and grain boundaries in single-layer graphene atomic patchwork quilts. Nature. 2011;**469**:389-392

[69] Rasool HI, Ophus C, Klug WS, Zettl A, Gimzewski JK. Measurement of the intrinsic strength of crystalline and polycrystalline graphene. Nature Communications. 2013;**4**: 2811. DOI: 10.1038/ncomms3811

[70] Lee C, Wei X, Kysar JW, Hone J. Measurement of the elastic properties and intrinsic strength of monolayer graphene. Science. 2008;**321**:385-388. DOI: 10.1126/science.1157996

[71] Tuan DV, Kotakoski J, Louvet T, Ortmann F, Meyer JC, Roche S. Scaling properties of charge transport in polycrystalline graphene. Nano Letters. 2013;**13**:1730-1735. DOI: 10.1021/nl400321r

[72] Zhang H, Lee G, Gong C, Colombo L, Cho K. Grain boundary effect on electrical transport properties of graphene. Journal of Physical Chemistry C. 2014;**118**:2338-2343. DOI: 10.1021/jp411464w

[73] Boukhvalov DW, Katsnelson MI. Chemical functionalization of graphene with defects. Nano Letters. 2008;**8**:4373-4379. DOI: 10.1021/nl802234n

[74] Cummings AW, Duong DL, Nguyen VL, Tuan DV, Kotakoski J, Vargas JEB, et al. Charge transport in polycrystalline graphene: Challenges and opportunities. Advanced Materials. 2014;**26**:5079-5094. DOI: 10.1002/adma.201401389

[75] Chen JH, Li L, Cullen WG, Williams ED, Fuhrer MS. Tunable Kondo effect in graphene with defects. Nature Physics. 2011;**7**:535-538. DOI: 10.1038/nphys1962

[76] Gogotsi Y. Nanomaterials Handbook. Boca Raton, Florida, United States: CRC Press; 2006

[77] Blocher JM, Browning MF, Barrett DM. Emergent Process Methods for High-Technology. Berlin, Germany: Ceramics. Springer; 1984. pp. 299-316

[78] Morosanu CE. Thin Films by Chemical Vapour Deposition. Elsevier; 1990. pp. 1-717

[79] Bhaviripudi S, Jia X, Dresselhaus MS, Kong J. Role of kinetic factors in chemical vapor deposition synthesis of uniform large area graphene using copper catalyst. Nano Letters. 2010;**10**:4128-4133. DOI: 10.1021/nl102355e

[80] Miao C, Zheng C, Liang O, Xie YH. Physics and Applications of Graphene-Experiments, Book Chapter. Chemical Vapor Deposition of Graphene. US: InTech; 2011

[81] Bae S, Kim H, Lee Y, Xu X, Park JS, Zheng Y, et al. Roll-to-roll production of 30-inch graphene films for transparent electrodes. Nature Nanotechnology 2010;**5**:574-578. DOI: 10.1038/nnano.2010.132

[82] Yu Q, Lian J, Siriponglert S, Li H, Chen YP, Pei SS. Graphene segregated on Ni surfaces and transferred to insulators. Applied Physics Letters. 2008;**93**:113103. DOI: 10.1063/1.2982585

[83] Kobayashi T, Bando M, Kimura N, Shimizu K, Kadono K, Umezu N, et al. Production of a 100-m-long high-quality graphene transparent film by roll-to-roll chemical vapor deposition and transfer process. Applied Physics Letters. 2013;**102**:023112. DOI: 10.1063/1.4776707

[84] Boyd DA, Lin WH, Hsu CC, Teague ML, Chen CC, Lo YY, et al. Single step deposition of high-mobility graphene at reduced temperatures. Nature Communications. 2015;**6**:6620. DOI: 10.1038/ncomms7620

[85] Jang J, Son M, Chung S, Kim K, Cho C, Hun LB, et al. Low-temperature-grown continuous graphene films from benzene by chemical vapor deposition at ambient pressure. Scientific Reports. 2015;**5**:17955. DOI: 10.1038/srep17955

[86] Gong Y, Zhang X, Liu G, Wu L, Geng X, Long M, et al. Layer-controlled and wafer-scale synthesis of uniform and high-quality graphene films on a polycrystalline nickel catalyst. Advanced Functional Materials. 2012;**22**:3153-3159. DOI: 10.1002/adfm.201200388

[87] Tu Z, Liu Z, Li Y, Yang F, Zhang L, Zhao Z, et al. Controllable growth of 1-7 layers of graphene by chemical vapour deposition. Carbon. 2014;**73**:252-258. DOI: 10.1016/j.carbon.2014.02.061

[88] Suk JW, Kitt A, Magnuson CW, Hao Y, Ahmed S, An J, et al. Transfer of CVD-grown monolayer graphene onto arbitrary substrates. ACS Nano. 2011;**5**:6916-6924. DOI: 10.1021/nn201207c

[89] Yan Z, Lin J, Peng Z, Sun Z, Zhu Y, Li L, et al. Toward the synthesis of wafer-scale single-crystal graphene on copper foils. ACS Nano. 2012;**6**:9110-9117. DOI: 10.1021/nn303352k

[90] Barcenas AM, Robles JFP, Vorobiev YV, Soto NO, Mexicano A, Garcia AG. Graphene synthesis using a CVD reactor and a discontinuous feed of gas precursor at atmospheric pressure. Journal of Nanomaterials. 2018;**2018**:1-11. DOI: 10.1155/2018/3457263

[91] Wang S, Hibino H, Suzuki S, Yamamoto H. Atmospheric pressure chemical vapor deposition growth of millimeter-scale single graphene on the copper surface with a native oxide layer. Chemistry of Materials. 2016;**28**:4893-4900. DOI: 10.1021/acs.chemmater.6b00252

[92] Campos-Delgado J, Botello-Mendez AR, Algara-Siller G, Hackens B, Pardoen T, Kaiser U, et al. CVD synthesis of mono and few-layer graphene using alcohols at low hydrogen concentration and atmospheric pressure. Chemical Physics Letters. 2013;**584**:142-146. DOI: 10.1016/j.cplett.2013.08.031

[93] Xu X, Zhang Z, Qiu L, Zhuang J, Zhang L, Wang H, et al. Ultrafast growth of single-crystal graphene assisted by a continuous oxygen supply. Nature Nanotechnology. 2016; **11**:930-935. DOI: 10.1038/nnano.2016.132

[94] Wu X, Zhong G, D'Arsie L, Sugime H, Esconjauregui S, Robertson AW, et al. Growth of continuous monolayer graphene with millimeter-sized domains using industrially safe conditions. Scientific Reports. 2016;**6**:21152. DOI: 10.1038/srep21152

[95] Mohsin A, Liu L, Liu P, Deng W, Ivanov IN, Li G, et al. Synthesis of millimeter-size hexagon-shaped graphene single crystals on resolidified copper. ACS Nano. 2013;**7**:8924-8931. DOI: 10.1021/nn4034019

[96] Hess DW, Jensen KF, Anderson TJ. Chemical Vapor Deposition. 2011. DOI: 10.1515/revce.1985.3.2.97

[97] Ishizaka A, Shiraki Y. Low-temperature surface cleaning of silicon and itsapplication to silicon MBE. Journal of the Electrochemical Society. 1986;**133**:666-671. DOI: 10.1149/1.2108651

[98] Kidambi PR, Ducati C, Dlubak B, Gardiner D, Weatherup RS, Martin MB, et al. The parameter space of graphene chemical vapor deposition on polycrystalline Cu. Journal of Physical Chemistry C. 2012;**116**:22492-22501. DOI: 10.1021/jp303597m

[99] Sun H, Han Y, Wu J, Lu Y, Xu J, Luo Y, et al. Cooling growth of milimeter-size single0-crystal bilayer graphene at atmospheric pressure. Journal of Physical Chemistry C. 2016; **120**:13596-13603. DOI: 10.1021/acs.jpcc.6b04105

[100] Qaisi RM, Smith CE, Hussain MM. Atmospheric pressure chemical vapor deposition (APCVD) grown bi-layer graphene transistor characteristics at high temperature. Physica Status Solidi RRL: Rapid Research Letters. 2014;**8**:621-624. DOI: 10.1002/pssr.201409100

Section 4

CVD of Nanofibers

Chapter 7

Chemical Vapor Deposition of Helical Carbon Nanofibers

Yoshiyuki Suda

Additional information is available at the end of the chapter

http://dx.doi.org/10.5772/intechopen.81676

Abstract

Helical carbon nanofibers (HCNFs), including carbon nanocoils (CNCs), carbon nanotwists, and multi-walled CNCs, can be synthesized by chemical vapor deposition (CVD). HCNFs are predicted to have a high mechanical strength and hence are expected to have a use in nanodevices such as electromagnetic wave absorbers and sensors. For nanodevice applications, it is necessary to synthesize HCNFs in high yield and purity. In this chapter, I focus on CNCs and describe its history, expected application, and synthesis method. Finally, I introduce the author's recent studies on the improvement of the purity of CNCs by improving CVD conditions. A CNC layer with a thickness of larger than 40 μm was grown from a triple layer of $SnO_2/Fe_2O_3/SnO_2$ catalyst on a substrate, and the CNC purity increased to 81 ± 2%.

Keywords: helical carbon nanofibers, carbon nanocoils, electromagnetic wave absorber, sensing devices, energy devices, binary catalyst, tetramethyltin, ferrocene, iron oxide fine powder

1. Introduction

1.1. Helical carbon nanofibers and carbon nanocoils

Carbon is an element with various allotropes because of its variety of chemical bonding: sp^1, sp^2, and sp^3. If we classify each carbon nanomaterial by its dimension, fullerene is a zero-dimensional substance, carbon nanotube (CNT) is a one-dimensional substance, and graphene is a two-dimensional substance. Helical carbon nanofiber (HCNF) is a helical-shaped carbon nanofiber with coil and fiber diameters in the range of 20–1000 nm and 5–400 nm, respectively. HCNFs are recognized as a three-dimensional substance according to the above definition.

There are two types of HCNFs: the first type, called CNCs, has an inner diameter and the second type does not have an inner diameter, and it is called carbon nanotwists. CNCs are classified into two types by their size and crystalline structure: CNCs and multi-walled CNCs (or coiled CNTs) which are the multi-walled CNTs with a coil geometry. In this chapter, I focus on CNCs (**Figure 1(a)**).

1.2. History

As far as I know, Davis et al. are the group who discovered CNCs. They have found minute vermicular growths of carbon through the experimental work on the deposition of carbon in the brickwork of blast furnaces in 1953 [1]. Two years later from the discovery, Hofer et al. reported the growth of carbon filaments with fiber diameters of 10–200 nm. They also found that the filaments were extended in two directions from central catalyst particles [2]. They used a catalytic chemical vapor deposition (CVD) technique of carbon from carbon monoxide.

After a while, Baker et al. developed a catalytic CVD using C_2H_2 gas over catalyst surfaces and studied the effect of Sn on the growth of filaments from the Fe/C_2H_2 system. They found that almost all the filaments had a spiral shape with a constant pitch and the Fe/Sn alloy worked to form spiral filaments [3, 4]. At the same time as Bakers' study, Boehm developed a continuous process of carbon filaments by feeding metal carbonyl into a CO stream before it passed a heated tube. He found that double or even triple helices were formed in the twisted filaments [5].

Carbon nanomaterials have been discovered one after another since the 1980s. Although fullerene [6], CNT [7], and carbon nanohorn (CNH) [8] were firstly synthesized by an arc discharge, CVD became a dominant technique in the research of carbon nanomaterials including HCNFs. In 1990, Motojima et al. have grown regularly coiled carbon filaments with a coil diameter of several micrometer by a catalytic CVD at 350–750°C using C_2H_2 gas feedstock and Ni as a catalyst [9]. Concurrently, Ivanov et al. have grown multi-walled CNC (or coiled CNT) which has a hollow structure with several concentric graphitic layers [10]. The CVD of single-walled CNTs (SWCNTs) was successfully realized in 1998 [11].

It is commonly accepted in CVD process that hydrocarbon precursors dissociate on catalyst particles, and the formed carbon species dissolve into catalyst particles and recombine to form carbon filaments. Models have been proposed to explain the formation of a helical structure in CNCs. Baker et al. postulated a bulk-diffusion model of carbon through catalyst particles based on the temperature gradient through a catalyst particle [3]. Another model was proposed by Amelinckx et al. and Fonseca et al. [12, 13]. The mismatch between extrusion velocities of different faces of a catalyst particle results in the deformation of carbon deposits.

1.3. Expected applications

Because of the structure of a solenoid coil, the application of electromagnetic wave absorber using CNCs [14–18] has been studied for the longest as an application of CNC. This is an application as a structural material with excellent electromagnetic wave absorption characteristics by mixing CNC with resin. It is comparatively easy and feasible for CNC application.

http://dx.doi.org/10.5772/intechopen.81676

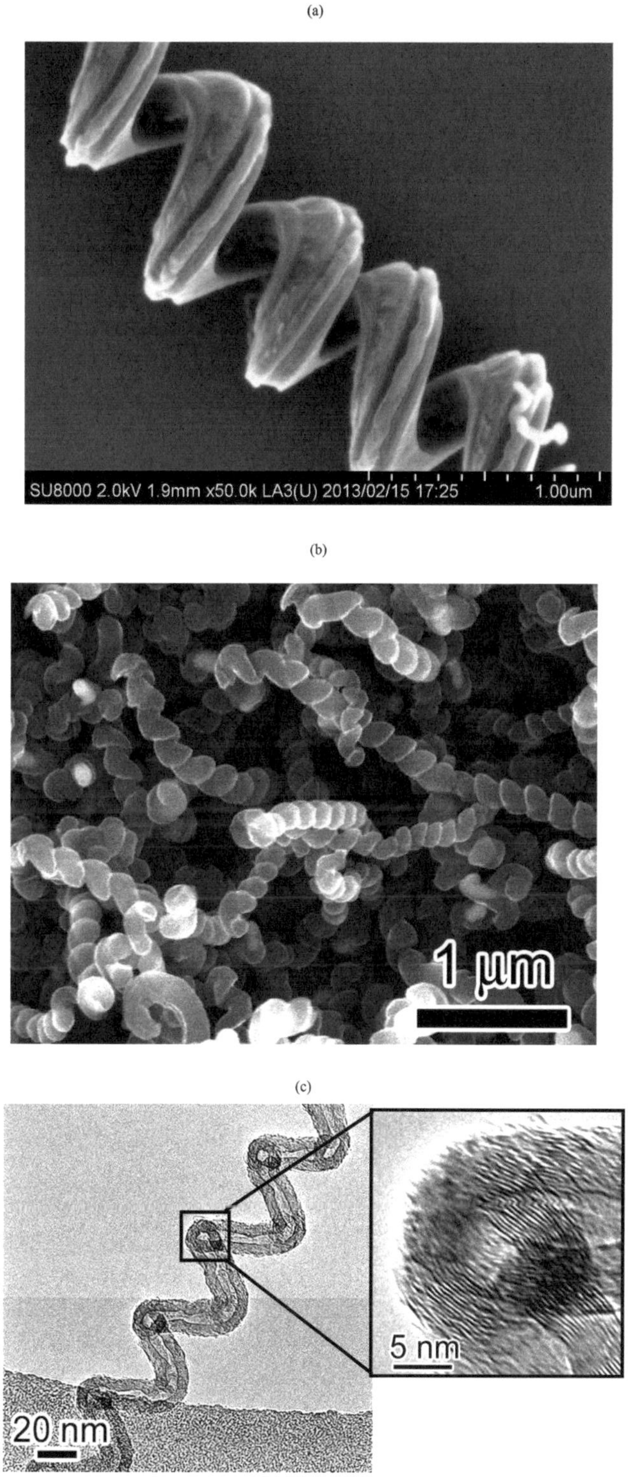

Figure 1. Scanning electron micrograph of HCNFs: (a) CNC, (b) carbon nanotwist, and (c) multi-walled CNC.

As a similar example, there is a fabrication of a sensing device in which plural CNCs are randomly integrated [19]. CNC-based biosensors [19], thermal sensors [20], and pressure and strain sensors [21–23] are reported one after another, and further development is expected.

On the other hand, there is no movement to use a single CNC as a nanospring or a nanoinductor, and it remains within the scope of basic research that it measures the spring constant [24] and estimates the inductance by simulation [25]. This is considered to be attributable to the fact that it takes a lot of trouble to arrange a single CNC in a desired place and direction and suitable application is not developed.

I am currently paying attention to the application of CNC to energy devices such as fuel cells [26–32] and electric double layer capacitors [33–37]. In this field, although carbon black and activated carbon, whose mass-production processes have been established, are used as a major carbon material, the applications of carbon nanomaterials including CNCs are progressing little by little. It is difficult at present to produce CNC at the same cost as these materials; its electrochemical properties higher than existing materials are however reported at the laboratory level. Future application of CNC is greatly expected.

2. Growth method

As fullerenes and CNTs were initially synthesized in high density plasmas such as laser vaporization and arc discharge, plasma processes have been used for synthesis of carbon nanomaterials. Meanwhile, CVD method is the mainstream of CNC since it was introduced in the 1970s. From the viewpoint of the carbon species to be supplied as a raw material, the method using plasma and the CVD method are summarized; a major difference is in between atomic and/or molecular carbon and hydrocarbon molecules.

Most hydrocarbons and alcohols are gaseous or liquid at ordinary temperature and normal pressure and can be easily introduced into the reaction furnace by using differential pressure or carrier gas. The hydrocarbon molecules supplied into the reactor are thermally decomposed by the catalytic reaction to change to the desired carbon nanomaterial. This process is called CVD method. Since synthesis of CNC by CVD method has much in common with synthesis of CNT, synthesis of CNT will be described first. As common to both materials, it is very important to arrange the supply of carbon feedstock and the synthesis site: catalyst and substrate.

In 1998, Dai's group succeeded in synthesizing single-walled CNTs by fixing Fe_2O_3 nanoparticles on a substrate and using CH_4 gas as a raw material [11]. This is explained by that hydrocarbon molecules are decomposed on the surface of Fe nanoparticles reduced in a high-temperature CH_4 atmosphere, carbon dissolves inside the nanoparticles (or diffuses on the surface), and carbon segregates from the nanoparticles to form CNT. This concept is basically the same as the synthesis process of carbon fiber, which was known before the discovery of CNTs [38]. Dai's group realized the growth of single-walled CNTs by maintaining the size of Fe nanoparticles to a minimum. On the other hand, in the synthesis of graphene, Cu in which the solubility of carbon is low is used as a catalyst. In this way, new catalysts have been developed.

Now, in order to synthesize CNC, the element which does not dissolve carbon inside such as Sn is mixed with Fe. The author tried various binary catalysts and showed that CNC can be synthesized from Fe/Sn catalyst with high efficiency [39]. Pan et al. synthesized CNC with high efficiency using a mixed catalyst of Fe and indium tin oxide (ITO) and proposed a model that the coil shape is formed due to the nonuniformity of the segregation speed of carbon from the catalyst nanoparticles [40]. There are several catalysts suitable for CNC growth, and there is a possibility that new combinations can be found in the future.

3. Recent improvement on CNC growth technique

It is now possible to produce 2–3 g of HCNFs on a catalyst-supported graphite substrate within an hour, using C_2H_2 feedstock [39]. However, problems of low CNC purity remain. The main problem of CNC synthesis is that CNCs can grow with high purity on the surface of carbon deposits, but interlayers with very low CNC purity are present inside the carbon deposit [39, 41, 42]. Here I introduce two techniques to improve CNC purity in CVD and the evaluation method of CNC purity. The CNC purity in most reports was evaluated by observing the surface of the carbon deposit. The variations in the CNC purity inside carbon deposits have not been yet understood completely.

3.1. Supplying catalyst molecules from the vapor phase

In this study, the Sn/Fe catalysts were supplied using the following four materials: a thin Sn film, a drop-coated solution of Fe_2O_3, tetramethyltin (TMT) vapor, and ferrocene vapor. The CNC purity averaged over the overall carbon deposit was increased 1.5-fold by the TMT supply. A maximum CNC purity of 72% using a combination of TMT and ferrocene vapors was obtained, with Sn/Fe deposition on the substrate [43].

Various experimental methods have been proposed for the production of high-purity CNCs [44–47]. However, the supply of vapor-phase catalysts during the synthesis of CNCs has been reported in only a few previous studies [48, 49]. In this study, we performed experiments in which different catalyst metal vapors were supplied successively in a CVD reactor.

3.1.1. Experimental setup and procedure

The experiments were performed in a horizontal CVD reactor using the following steps: (1) vacuum-evaporating Sn on the SiO_2/Si substrate, (2) drop-coating an Fe_2O_3 solution on the Sn-coated SiO_2/Si substrate and performing calcination, (3) mounting the Sn/Fe-coated SiO_2/Si substrate in the CVD reactor, (4) performing CVD using a C_2H_2/N_2 gas mixture, and (5) investigating the effects of changing the catalytic vapor supply, adding tetramethyltin (TMT) and/or ferrocene to C_2H_2/N_2 gas mixture during CVD.

Sn and Fe vapor catalysts were introduced into the CVD reactor using a tube pump and a vaporizer. TMT ($Sn(CH_3)_4$, purity: >96%, CAS No.: 594–27-4, Tokyo Chemical Industry Co., Ltd., Tokyo, Japan) and ferrocene ($Fe(C_5H_5)_2$ purity: > 95%, CAS No.: 102–54-5, Tokyo

Chemical Industry Co., Ltd., Tokyo, Japan) were used as the feedstocks for the Sn and Fe catalytic vapors, respectively. These catalyst molecules were mixed in ethanol, supplied to a vaporizer via a tube pump, and vaporized in a vaporizer. The CVD experimental conditions and catalytic vapor supply conditions are listed in **Tables 1** and **2**.

The synthesized CNCs and the catalyst nanoparticles were observed using SEM (S-4500II and SU8000, Hitachi High-Technologies Corp., Tokyo, Japan). The CNC purity was evaluated by analyzing SEM images of (1) the top surface and (2) the cross section of the samples. The carbon deposit was solidified by dropping a Nafion® solution onto the samples; then, the carbon deposit was cut to observe the cross section.

3.1.2. Results and discussion

In the experiment, the Fe catalyst was formed on the substrate before CVD and TMT vapors were supplied during CVD. The authors confirmed the growth of CNCs in all of the TMT concentration conditions (0.1–2% in ethanol) used in this study. The CNC number density was evaluated from the SEM images obtained at each TMT concentration, as shown in **Figure 2**. The CNC number density was highest at a TMT concentration of 0.2%. As the TMT concentration was higher than 0.2%, the CNC number density tended to decrease. The amount of carbon deposits showed the same tendency as the CNC number density.

By the above experimental results, the TMT concentration in ethanol was fixed at 0.2%, and the thickness of the Sn thin film was changed. CNC grew with all the Sn film thicknesses. Particularly, high-purity CNCs were obtained with the thicknesses of 40 and 60 nm. The CNC purity reached maximum (about 62%) for an Sn film thickness of 40 nm.

Figure 3 shows the as-grown carbon deposits on the substrate with a Sn film thickness of 40 nm and SEM micrographs of their cross sections. It was shown that the CNC purity inside

Flow rates of C_2H_2/N_2 gases	50/1000 mL/min
Total gas pressure	1.013×10^5 Pa
Synthesis temperature	700°C
Growth time	30–600 s
Substrate	SiO_2/Si
Sn film thickness on substrate	0–60 nm
Amount of Fe_2O_3 solution (3% concentration) dropped on substrate	10 μL

Table 1. CVD conditions.

TMT concentration in ethanol	0.1–2%
Flow rate of TMT solution	0.144 mL/min
Temperature inside the carburetor	80°C

Table 2. Catalytic vapor supply conditions.

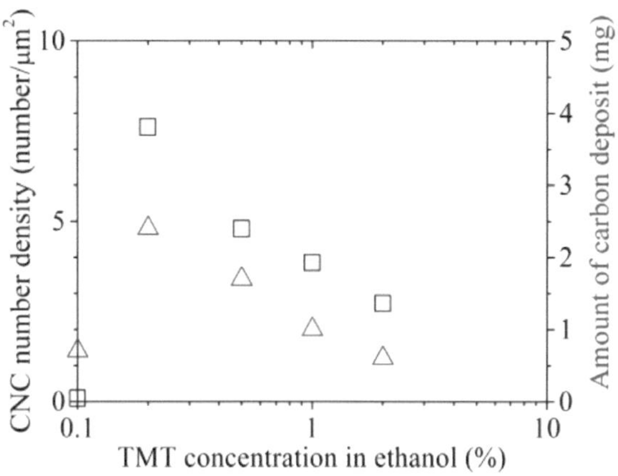

Figure 2. Dependence of the CNC number density (open squares) and amount of carbon deposits (open triangles) on the TMT concentration in ethanol. No Sn film was deposited on the SiO_2/Si substrates. The growth time was 600 s. The other growth conditions were the same as described in **Tables 1** and **2**.

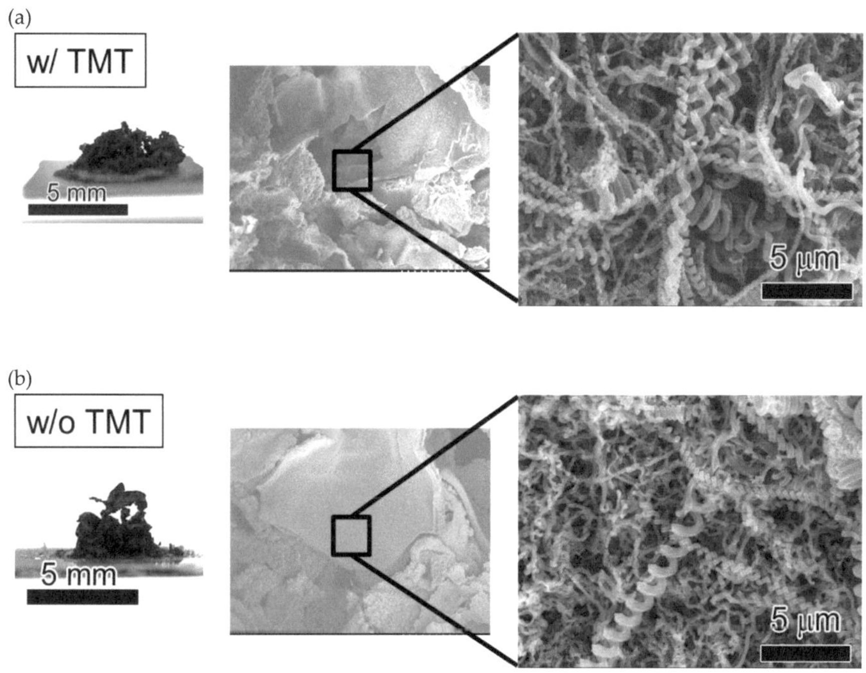

Figure 3. Photographs of as-grown carbon deposits and SEM micrographs of their cross sections: (a) with and (b) without TMT. The TMT concentration in ethanol, Sn film thickness, and growth time were 0.2%, 40 nm, and 600 s, respectively. The other growth conditions were the same as described in **Tables 1** and **2**.

the carbon deposit was lower than that of the surface from the SEM observation, regardless of TMT supply. The average CNC purities (average over the total carbon deposit) evaluated from the SEM image of the cross section of the carbon deposit were 42% for TMT and 28% without TMT. The CNC purity increased by 1.5 times by TMT supply. As shown in **Figure 4**, by the energy dispersive spectroscopy (EDS) analysis of the cross section of the carbon deposit,

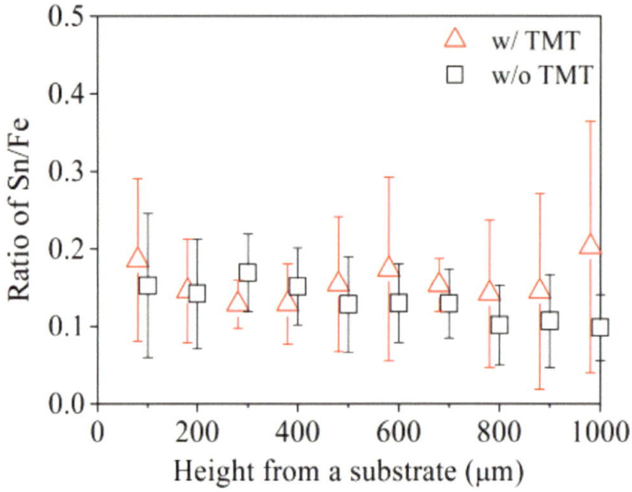

Figure 4. Relationship between the Sn/Fe ratio at the cross-sectional surfaces of the carbon deposits and the height from the substrate. The TMT concentration in ethanol, Sn film thickness, and growth time were 0.2%, 40 nm, and 600 s, respectively. The other growth conditions were the same as described in **Tables 1** and **2**.

every 100 μm from the substrate surface to a height of 1000 μm revealed that the molar ratio of Sn to Fe (Sn/Fe ratio) obtained with TMT supply was always higher than those without TMT supply. This suggests that Sn catalyst could be supplied by TMT to the interior of the carbon deposit as well as the substrate surface.

Next, the result of adding ferrocene to the TMT solution is explained. The molar ratio of Fe: Sn in the solution containing ferrocene was set to 3: 1. As shown in **Figure 5**, the CNC purity on the surface of carbon deposits obtained with different Sn film thicknesses was evaluated. It was clearly shown that the supply of ferrocene increases CNC purity. When the Sn film thickness was 40 nm, the purity of CNC reached a maximum (72%).

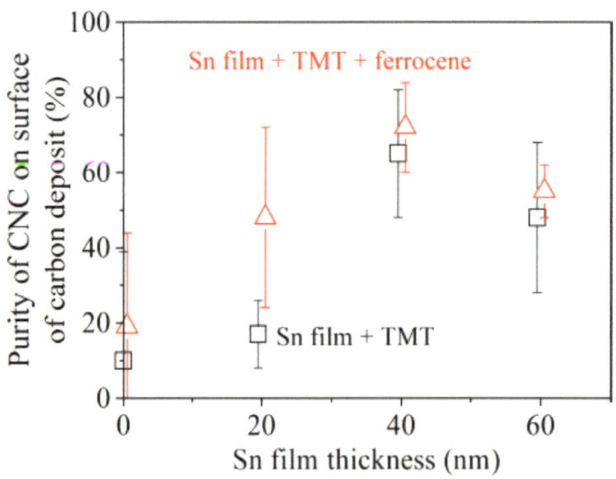

Figure 5. Purity of CNCs grown on the surface of carbon deposits under a supply of TMT and ferrocene. The TMT concentration in ethanol and growth time was 0.2% and 600 s, respectively. The other growth conditions were the same as described in **Tables 1** and **2**.

When ferrocene is added, not only Sn but also Fe is supplied as steam. Then, it is possible that these catalysts will grow CNC in the gas phase. In order to confirm this, carbon deposits on the inner wall of the CVD apparatus were analyzed after the CVD experiment. Several CNFs were observed by SEM, but the presence of CNC was not confirmed. Since ferrocene and TMT could be decomposed at the CVD growth temperature used in this study [50, 51], it seems reasonable that CNFs were grown from catalyst nanoparticles in the vapor phase. Li et al. however argue that it is necessary for the CNC catalyst nanoparticles to be fixed on a substrate to form the CNCs' helical structure [41].

3.2. Using iron oxide fine powder as a catalyst

In this section, the experimental results that increase the purity of CNC by improving the structure of the catalyst layer and CVD conditions are introduced [52]. By forming a triple-layer structure of $SnO_2/Fe_2O_3/SnO_2$ catalyst on the substrate and optimizing the CVD conditions, a CNC layer with a thickness of 40 μm or more was grown, and the CNC purity inside this layer successfully raised to 81 ± 2%.

There is a carbon layer that does not contain CNCs between the upper and lower CNC layers; this was called the carbon layer. Both the CNC and carbon layers were grown from a catalyst that was deposited on the substrate. In the author's experiment, when this carbon layer was taken into account, the CNC purity over the whole deposits grown became almost 55%, very low compared to the purity of the CNC layer. Yokota et al. [39] reported that the thickness of the carbon layer depends on the thickness of the catalyst film on the substrate.

In order to increase the CNC purity and reduce the thickness of the carbon layer, formation of the catalyst layer by using a spin coating method and Fe fine powder was attempted. CNCs were synthesized using an automatic CVD apparatus.

3.2.1. Experimental setup and procedure

The CVD conditions in this study are listed in **Table 3**. The detail of the automatic CVD apparatus is described elsewhere [53].

In this experiment, the authors replaced a Fe powder (diameter: 1–3 μm) with a Fe fine powder (diameter: 20 nm) and used the spin coating method for coating Fe on the substrate in order to reduce the catalyst film thickness. Solution of Fe_2O_3 (0.1 M) and SnO_2 (0.13 M) was separately coated on the Si substrate. The spin coating speed was varied between 1000 rpm and 2500 rpm. After coating one kind of solution, the substrate was dried at 80°C for 5 minutes to suppress mixture of the solutions. The following three kinds of catalyst structures were formed: (a) Fe_2O_3/SnO_2, (b) SnO_2/Fe_2O_3, and (c) $SnO_2/Fe_2O_3/SnO_2$. By observing the catalyst structure with SEM, the optimum catalyst structure for CNC growth and the influence of this structure on the thickness of the carbon layer were investigated [52].

The purity of CNCs over the entire deposit was evaluated. As illustrated in **Figure 6**, the carbon deposit is sheet-like in shape, and CNCs were grown on both sides of the carbon layer.

Catalyst precursor	Fe_2O_3 fine powder (diameter: 20 nm, Nilaco, Tokyo, Japan) SnO_2 drop-coating solution (0.13 M, Kojundo Chemical Laboratory, Sakado, Japan)
Spin coating velocity	1000–2500 rpm
Catalyst structure	Fe_2O_3/SnO_2, $SnO_2/Fe_2O_3/SnO_2$, SnO_2/Fe_2O_3
Catalyst molar ratio	Fe:Sn = 1:2.6–1:13
Feedstock gas (flow rate)	C_2H_2 (100–400 mL/min)
Dilution gas (flow rate)	N_2 (1000–1800 mL/min)
Synthesis temperature	780°C
Synthesis time	0.5–30 min
Annealing temperature	780°C
Annealing time	5 min

Table 3. Catalytic CVD conditions.

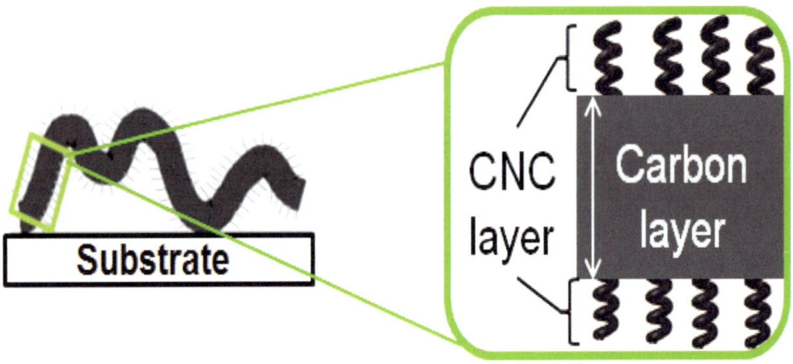

Figure 6. Illustration of the carbon deposit.

The CNC purity was defined by Eq. (1) under the assumption that the CNC purities in the CNC and the carbon layers are 100% and 0%, respectively.

$$\text{CNC purity} = \frac{\text{Thickness of CNC layer}}{\text{Thicknesses of CNC and carbon layers}} \times 100\ (\%) \tag{1}$$

The assumption is consistent with the SEM observations of the deposits. The thickness of the CNC layer was obtained as the summation of the upper and lower layers.

3.2.2. Results and discussion

Firstly, the effect of flow rates of C_2H_2 feedstock and N_2 dilution gases on CNC growth was examined. The conditions in this experiment are as follows: catalyst structure, triple layer ($SnO_2/Fe_2O_3/SnO_2$); Fe/Sn molar ratio, 1:2.6; spin coating velocity, 1000 rpm; and synthesis time, 10 min. The other conditions are the same as the conditions shown in **Table 2**. In the experiment in which the gas flow rate was varied, the flow rate of C_2H_2 gas was fixed at 250 mL/min, and the flow rate of N_2 gas was varied. Similarly, the flow rate of C_2H_2 gas was changed by fixing the flow rate of N_2 gas to 1400 mL/min.

The thickness ratio of the CNC layer and the carbon layer was examined by changing the flow ratio between C_2H_2 gas and N_2 gas. As a result, the thickness of the CNC layer as well as that of the carbon layer was maximized with a C_2H_2/N_2 gas flow ratio of 0.18 and decreased when the C_2H_2/N_2 gas flow ratio was greater than 0.18. Also, the difference in thickness between the CNC and the carbon layer was maximized at a C_2H_2/N_2 gas flow ratio of 0.18. When the C_2H_2/N_2 gas flow rate ratio was in the range of 0.15–0.25, the purity of CNC in the CNC layer was 95% or more.

Then, the influence of the spin coating velocity was investigated. The experimental conditions different from the experiment of the gas flow rate dependence are as follows: catalyst structure on substrate, two layers (Fe_2O_3/SnO_2); C_2H_2 gas flow rate, 250 mL/min; and N_2 gas flow rate, 1400 mL/min. Other conditions are the same as those shown in **Table 2**. The thicknesses of the CNC layer and the carbon layer with respect to the spin coating velocity are shown in **Figure 7**. The thickness of the carbon layer decreased with increasing spin coating velocity. It is believed that the amount of liquid remained on the substrate was reduced by increasing the spin coating velocity from 1000 to 2500 rpm. Since the CNC layer was hardly formed at a spin coating rate of 1500 rpm or more, the upper limit of the spin coating velocity seems to be 1000 rpm. From the experimental results, the optimum spin coating velocity was 1000 rpm. At this velocity, the CNC purity became highest, but the thickness of the carbon layer was also highest.

The effect of synthesis time on CNC growth was also examined. As synthesis time increased from 30 s to 10 min, the CNC purity increased constantly. However, the CNC purity remained almost constant at 75% for synthesis times between 10 and 30 min, indicating saturation tendency.

Next, the influence of the catalyst structure on carbon deposition was investigated using three types of catalyst structures. Different conditions from the above two experiments are as follows: spin coating velocity, 1000 rpm and synthesis time, 30 min. Other conditions are the

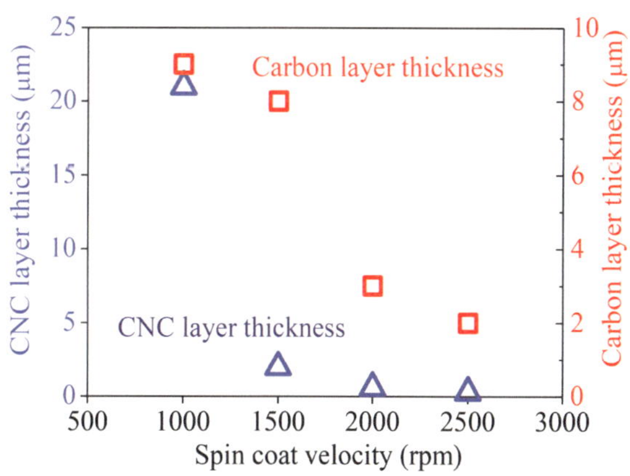

Figure 7. Thicknesses of CNC and carbon layers vs. spin coating velocity.

same as those listed in **Table 2**. Cross-sectional SEM micrographs of carbon deposits containing CNC and carbon layers grown from three types of catalyst structures and SEM micrographs of the surface of the CNC layers are shown in **Figures 8** and **9**, respectively.

As shown in **Figure 8**, a CNC layer was formed on the catalyst structures of (a) Fe_2O_3/SnO_2 and (b) $SnO_2/Fe_2O_3/SnO_2$. The purities of CNCs in (a) and (b) were estimated to be 69% ± 2% and 81% ± 2%, respectively, according to Eq. (1). The reason for that the CNC purity differs depending on the catalyst structure was examined using a CNC growth model as shown in **Figure 10**.

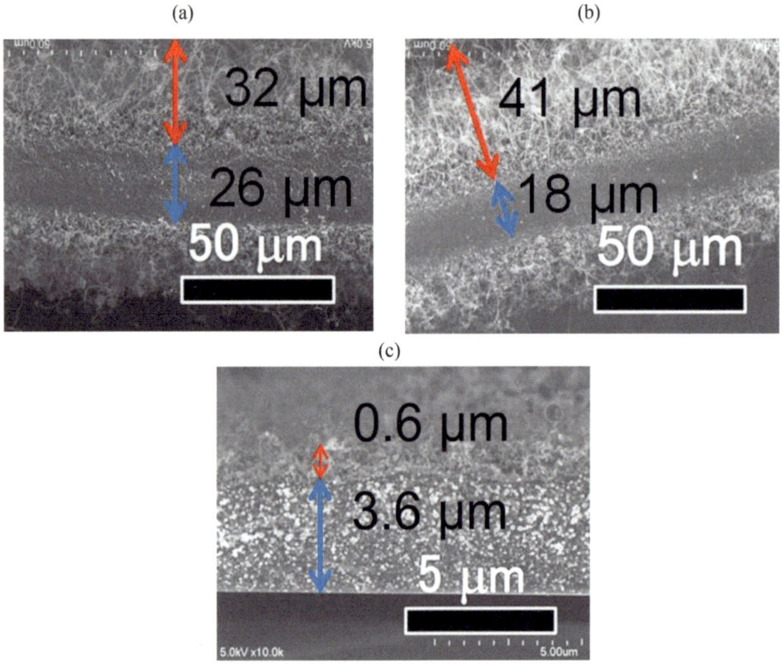

Figure 8. Cross-sectional SEM micrographs of carbon deposits grown with different catalyst structures: (a) Fe_2O_3/SnO_2, (b) $SnO_2/Fe_2O_3/SnO_2$, and (c) SnO_2/Fe_2O_3. The red and blue arrows represent the thicknesses of the CNC and carbon layers, respectively.

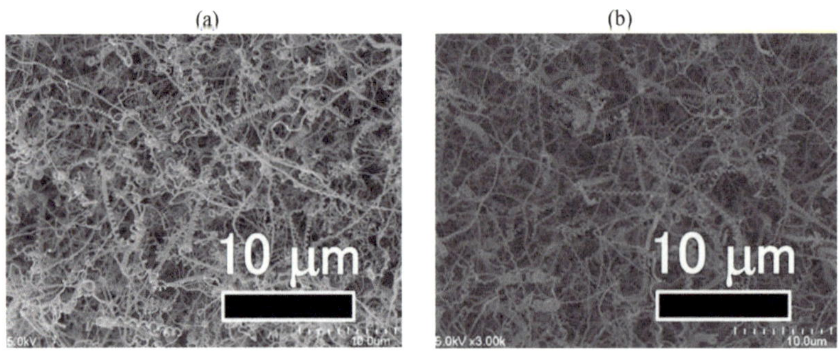

Figure 9. SEM micrographs of the surfaces of CNC layers grown from different catalyst structures: (a) Fe_2O_3/SnO_2 and (b) $SnO_2/Fe_2O_3/SnO_2$. No CNCs were observed on the SnO_2/Fe_2O_3 catalyst structure.

http://dx.doi.org/10.5772/intechopen.81676

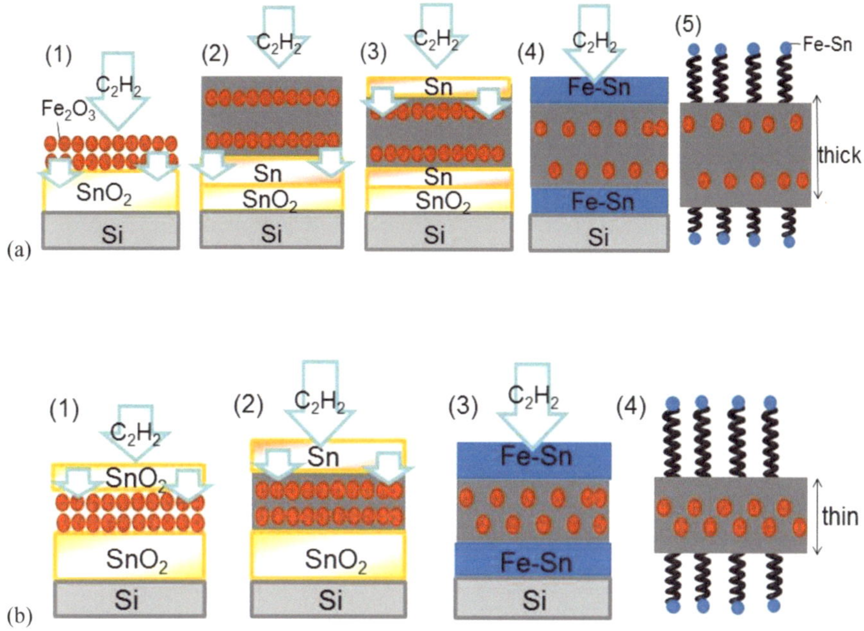

Figure 10. Growth models of (a) Fe_2O_3/SnO_2 and (b) $SnO_2/Fe_2O_3/SnO_2$ catalyst structures.

When C_2H_2 source gas molecule is supplied on the Fe_2O_3/SnO_2 catalyst, Fe_2O_3 in the catalyst forms nanoparticles ((1) in **Figure 10a**). A part of C_2H_2 molecule passes through the Fe_2O_3 to SnO_2, and SnO_2 is partially reduced by C_2H_2 molecules. Since the melting point of Sn is 232°C., which is much lower than the CVD temperature, a part of the reduced Sn stays as a liquid phase, and the other part diffuses inside the catalyst ((2) in **Figure 10a**). This phenomenon is explained by previous research results that molten Sn diffuses through the pores in the Fe_2O_3 thin film by capillary action [43]. However, since the amount of the Fe-Sn alloy formed on the surface of the Fe_2O_3/SnO_2 structure is limited ((3) in **Figure 10a**), a lot of sheets of carbon deposits are formed inside the Fe-Sn structure ((4) in **Figure 10a**). From the above, in the Fe_2O_3/SnO_2 catalyst structure, C_2H_2 molecules are directly supplied to Fe_2O_3 having a strong ability to decompose C_2H_2, and a very thick carbon layer is formed.

On the other hand, the amount of C_2H_2 molecules reaching the Fe_2O_3 catalyst nanoparticles in the $SnO_2/Fe_2O_3/SnO_2$ catalyst structure is limited ((1) in **Figure 10b**). By supplying more C_2H_2 molecules, an Fe-Sn alloy is formed on the upper and lower surfaces of $SnO_2/Fe_2O_3/SnO_2$, the same as in the case of the Fe_2O_3/SnO_2 catalyst structure ((3) in **Figure 10b**). However, the thickness of the layer is greater than the thickness of the layer formed from the Fe_2O_3/SnO_2 catalyst structure due to the much amount of Sn on the surface. Since CNC grows from the Fe-Sn alloy ((4) in **Figure 10b**), the thickness of the carbon layer grown inside the $SnO_2/Fe_2O_3/SnO_2$ catalyst structure becomes thinner than in the case of Fe_2O_3/SnO_2. This consideration explains the experimental results shown in **Figure 8**.

In the both models described above, the diffusion of C_2H_2 in the catalyst film is an important process for the growth of the CNC layer. This leads to a difference in thickness between the upper and lower CNC layers. As the experimental results show, the thickness of the upper CNC layer was always thicker than the thickness of the lower CNC layer. This also explains the reason for that CNC hardly grows on the SnO_2/Fe_2O_3 catalyst structure. This indicates that when the SnO_2 layer is too thick, the amount of C_2H_2 supplied to Fe_2O_3 is appreciably limited. Based on the above experimental results, the authors concluded that the optimum catalyst structure for growing CNC in high yield and high purity is the $SnO_2/Fe_2O_3/SnO_2$ catalyst structure.

4. Summary

CNC, one type of HCNFs, is introduced. CNC has a coil diameter of 20–1000 nm, and a length of several tens of μm and has mainly been synthesized by a CVD method. Although it is very small, CNC is predicted to have a high mechanical strength and an electrical property and hence is expected to have a use in nanodevices such as electromagnetic wave absorbers and material in energy devices. The history, expected application, and synthesis method were described. The authors' recent studies on the improvement of the purity of CNCs by improving CVD conditions were summarized.

Acknowledgements

This work was partly supported by the JSPS KAKENHI Grant Number 24360108 and 25630110; the Ministry of Education, Culture, Sports, Science and Technology (MEXT) KAKENHI Grant Number 24110708; and Analysis and Development System for Advanced Materials (ADAM), Research Institute for Sustainable Humanosphere (RISH), Kyoto University. I greatly appreciate the experimental contribution by Mr. Yuichi Ishii, Mr. Tatsuki Miki, and Mr. Koji Maruyama of Toyohashi University of Technology. This study was performed under the cooperative research with Tokai Carbon Co., Ltd.; Shonan Plastic Manufacturing Co., Ltd.; and Toho Gas Co., Ltd. I acknowledge Mr. Hitoshi Ue, Mr. Kazuki Shimizu, and Dr. Yoshito Umeda for their fruitful discussion.

Conflict of interest

The author declares that he has no conflict of interest.

Author details

Yoshiyuki Suda

Address all correspondence to: suda@ee.tut.ac.jp

Toyohashi University of Technology, Toyohashi, Japan

References

[1] Davis WR, Slawson RJ, Rigby GR. An unusual form of carbon. Nature. 1953;**171**:756. DOI: 10.1038/171756a0

[2] Hofer LJE, Sterling E, McCartney JT. Structure of carbon deposited from carbon monoxide on iron, cobalt and nickel. The Journal of Physical Chemistry. 1955;**59**:1153-1155. DOI: 10.1021/j150533a010

[3] Baker R. Nucleation and growth of carbon deposits from the nickel catalyzed decomposition of acetylene. Journal of Catalysis. 1972;**26**:51-62. DOI: 10.1016/0021-9517(72)90032-2

[4] Baker RTK, Harris PS, Terry S. Unique form of filamentous carbon. Nature. 1975;**253**:37-39. DOI: 10.1038/253037b0

[5] Boehm HP. Carbon from carbon monoxide disproportionation on nickel and iron catalysts: Morphological studies and possible growth mechanisms. Carbon. 1973;**11**:583-590. DOI: 10.1016/0008-6223(73)90323-0

[6] Kroto HW, Heath JR, O'Brien SC, Curl RF, Smalley RE. C_{60}: Buckminsterfullerene. Nature. 1985;**318**:162. DOI: 10.1038/318162a0

[7] Iijima S. Helical microtubules of graphitic carbon. Nature. 1991;**354**:56-58. DOI: 10.1038/354056a0

[8] Iijima S, Yudasaka M, Yamada R, Bandow S, Suenaga K, Kokai F, et al. Nano-aggregates of single-walled graphitic carbon nano-horns. Chemical Physics Letters. 1999;**309**:165-170. DOI: 10.1016/S0009-2614(99)00642-9

[9] Motojima S, Kawaguchi M, Nozaki K, Iwanaga H. Growth of regularly coiled carbon filaments by Ni catalyzed pyrolysis of acetylene, and their morphology and extension characteristics. Applied Physics Letters. 1990;**56**:321-323. DOI: 10.1063/1.102816

[10] Ivanov V, Nagy JB, Lambin P, Lucas A, Zhang XB, Zhang XF, et al. The study of carbon nanotubules produced by catalytic method. Chemical Physics Letters. 1994;**223**:329-335. DOI: 10.1016/0009-2614(94)00467-6

[11] Kong J, Cassell AM, Dai H. Chemical vapor deposition of methane for single-walled carbon nanotubes. Chemical Physics Letters. 1998;**292**:567-574. DOI: 10.1016/S0009-2614(98)00745-3

[12] Amelinckx S, Zhang XB, Bernaerts D, Zhang XF, Ivanov V, Nagy JB. A formation mechanism for catalytically grown helix-shaped graphite nanotubes. Science. 1994;**265**:635-639. DOI: 10.1126/science.265.5172.635

[13] Fonseca A, Hernadi K, Nagy Jb, Lambin P, Lucas AA. Model structure of perfectly graphitizable coiled carbon nanotubes. Carbon. 1995;**33**:1759-1775. DOI: 10.1016/0008-6223(95)00150-3

[14] Zhao D-L, Shen Z-M. Preparation and microwave absorption properties of carbon nanocoils. Materials Letters. 2008;**62**:3704-3706. DOI: 10.1016/j.matlet.2008.04.032

[15] Liu L, He P, Zhou K, Chen T. Microwave absorption properties of carbon fibers with carbon coils of different morphologies (double microcoils and single nanocoils) grown on them. Journal of Materials Science. 2014;**49**:4379-4386. DOI: 10.1007/s10853-014-8137-z

[16] Eguchi U, Takikawa H, Suda Y. Electromagnetic wave absorption characteristics of multiwalled carbon nanocoils. Japanese Journal of Applied Physics. 2014;**53**:045102. DOI: 10.7567/JJAP.53.045102

[17] Jian X, Chen X, Zhou Z, Li G, Jiang M, Xu X, et al. Remarkable improvement in microwave absorption by cloaking a micro-scaled tetrapod hollow with helical carbon nanofibers. Physical Chemistry Chemical Physics. 2015;**17**:3024-3031. DOI: 10.1039/C4CP04849K

[18] Li L, Lu L, Qi S. Preparation, characterization and microwave absorption properties of porous nickel ferrite hollow nanospheres/helical carbon nanotubes/polypyrrole nanowires composites. Journal of Materials Science: Materials in Electronics. 2018;**29**:8513-8522. DOI: 10.1007/s10854-018-8865-y

[19] Jia X, Hu G, Nitze F, Barzegar HR, Sharifi T, Tai C-W, et al. Synthesis of palladium/helical carbon nanofiber hybrid nanostructures and their application for hydrogen peroxide and glucose detection. ACS Applied Materials & Interfaces. 2013;**5**:12017-12022. DOI: 10.1021/am4037383

[20] Su C-C, Huang C-L, Chang S-H. Fabrication of aligned carbon nanocoil thermal sensor with a high temperature coefficient of electrical resistance at 25-100 °C. IEEE Transactions on Nanotechnology. 2015;**14**:794-797. DOI: 10.1109/TNANO.2015.2426611

[21] Li C, Pan L, Deng C, Wang P, Huang Y, Nasir H. A flexible, ultra-sensitive strain sensor based on carbon nanocoil network fabricated by an electrophoretic method. Nanoscale. 2017;**9**:9872-9878. DOI: 10.1039/C7NR01945A

[22] Deng C, Pan L, Zhang D, Li C, Nasir H. A super stretchable and sensitive strain sensor based on a carbon nanocoil network fabricated by a simple peeling-off approach. Nanoscale. 2017;**9**:16404-16411. DOI: 10.1039/C7NR05486F

[23] Li C, Pan L, Deng C, Cong T, Yin P, Wu Z. A highly sensitive and wide-range pressure sensor based on a carbon nanocoil network fabricated by an electrophoretic method. Journal of Materials Chemistry C. 2017;**5**:11892-11900. DOI: 10.1039/c7tc04166g

[24] Chen X, Zhang S, Dikin DA, Ding W, Ruoff RS, Pan L, et al. Mechanics of a carbon nanocoil. Nano Letters. 2003;**3**:1299-1304. DOI: 10.1021/nl0343670

[25] Wiselin J, Suseela SB, Jalaja BV, Ramani SDSP, Prasad R, Devaraj S, et al. A Low cost carbon nanofiber based spiral inductor: inference and implementation. Advances in Materials Science and Engineering. 2014;**2014**:1-8. DOI: 10.1155/2014/384917

[26] Park K-W, Sung Y-E, Han S, Yun Y, Hyeon T. Origin of the enhanced catalytic activity of carbon nanocoil-supported PtRu alloy electrocatalysts. The Journal of Physical Chemistry B. 2004;**108**:939-944. DOI: 10.1021/jp0368031

[27] Leela Mohana Reddy A, Jafri RI, Jha N, Ramaprabhu S, Ajayan PM. Carbon nanocoils for multi-functional energy applications. Journal of Materials Chemistry. 2011;**21**:16103. DOI: 10.1039/c1jm12580j

[28] Hu G, Nitze F, Sharifi T, Barzegar HR, Wågberg T. Self-assembled palladium nanocrystals on helical carbon nanofibers as enhanced electrocatalysts for electro-oxidation of small molecules. Journal of Materials Chemistry. 2012;**22**:8541. DOI: 10.1039/c2jm16075g

[29] Nitze F, Mazurkiewicz M, Malolepszy A, Mikolajczuk A, Kędzierzawski P, Tai C-W, et al. Synthesis of palladium nanoparticles decorated helical carbon nanofiber as highly active anodic catalyst for direct formic acid fuel cells. Electrochimica Acta. 2012;**63**:323-328. DOI: 10.1016/j.electacta.2011.12.104

[30] Suda Y, Shimizu Y, Ozaki M, Tanoue H, Takikawa H, Ue H, et al. Electrochemical properties of fuel cell catalysts loaded on carbon nanomaterials with different geometries. Materials Today Communications. 2015;**3**:96-103. DOI: 10.1016/j.mtcomm.2015.02.003

[31] Shimizu Y, Suda Y, Takikawa H, Ue H, Shimizu K, Umeda Y. Effective utilization of carbon nanocoil-supported PtRu anode catalyst by applying anode microporous layer for improved direct methanol fuel cell performance. Electrochemistry. 2015;**83**:381-385. DOI: 10.5796/electrochemistry.83.381

[32] Tang S, Huangfu H, Dai Z, Sui L, Zhu Z. Preparation of Fe-N-carbon nanocoils as catalyst for oxygen reduction reaction. International Journal of Electrochemical Science. 2015;**10**:12. http://www.electrochemsci.org/papers/vol10/100907180.pdf

[33] Rakhi RB, Cha D, Chen W, Alshareef HN. Electrochemical energy storage devices using electrodes incorporating carbon nanocoils and metal oxides nanoparticles. The Journal of Physical Chemistry C. 2011;**115**:14392-14399. DOI: 10.1021/jp202519e

[34] Rakhi RB, Chen W, Alshareef HN. Conducting polymer/carbon nanocoil composite electrodes for efficient supercapacitors. Journal of Materials Chemistry. 2012;**22**:5177. DOI: 10.1039/c2jm15740c

[35] Choi WH, Choi MJ, Bang JH. Nitrogen-doped carbon nanocoil array integrated on carbon nanofiber paper for supercapacitor electrodes. ACS Applied Materials & Interfaces. 2015;**7**:19370-19381. DOI: 10.1021/acsami.5b05527

[36] Izumi H, Okabe Y, Suda Y, Takikawa H, Tanoue H, Ue H, et al. Manufacturing of electric double-layer capacitors using carbon nanocoils and evaluation of their specific capacitance at a high scan rate. IEEJ Transactions on Fundamentals and Materials. 2013;**133**:660-667. DOI: 10.1541/ieejfms.133.660

[37] Yu L, Wan G, Qin Y, Wang C. Atomic layer deposition assisted fabrication of high-purity carbon nanocoil for electrochemical energy storage. Electrochimica Acta. 2018;**268**:283-294. DOI: 10.1016/j.electacta.2018.02.114

[38] Tibbetts GG. Carbon fibers produced by pyrolysis of natural gas in stainless steel tubes. Applied Physics Letters. 1983;**42**:666-668. DOI: 10.1063/1.94066

[39] Yokota M, Hosokawa Y, Shinohara Y, Kawabata T, Takimoto K, Suda Y, et al. Splitting and flattening of helical carbon nanofibers by acid treatment. Journal of Nanoscience and Nanotechnology. 2010;**10**:3910-3914. DOI: 10.1166/jnn.2010.1964

[40] Pan L, Zhang M, Nakayama Y. Growth mechanism of carbon nanocoils. Journal of Applied Physics. 2002;**91**:10058. DOI: 10.1063/1.1471575

[41] Li D, Pan L. Necessity of base fixation for helical growth of carbon nanocoils. Journal of Materials Research. 2012;**27**:431-439. DOI: 10.1557/jmr.2011.401

[42] Hirahara K, Nakayama Y. The effect of a tin oxide buffer layer for the high yield synthesis of carbon nanocoils. Carbon. 2013;**56**:264-270. DOI: 10.1016/j.carbon.2013.01.007

[43] Suda Y, Ishii Y, Miki T, Maruyama K, Tanoue H, Takikawa H, et al. Improvement of carbon nanocoil purity achieved by supplying catalyst molecules from the vapor phase in chemical vapor deposition. Journal of Materials Research. 2014;**29**:2179-2187. DOI: 10.1557/jmr.2014.247

[44] Chang N-K, Chang S-H. High-yield synthesis of carbon nanocoils on stainless steel. Carbon. 2008;**46**:1106-1109. DOI: 10.1016/j.carbon.2008.03.020

[45] Bi H, Kou K-C, Ostrikov KK, Wang Z-C. High-yield atmospheric-pressure CVD of highly-uniform carbon nanocoils using Co-P catalyst nanoparticles prepared by electroless plating. Journal of Alloys and Compounds. 2009;**484**:860-863. DOI: 10.1016/j.jallcom.2009.05.064

[46] Qi X, Zhong W, Deng Y, Au C, Du Y. Synthesis of helical carbon nanotubes, worm-like carbon nanotubes and nanocoils at 450°C and their magnetic properties. Carbon. 2010;**48**:365-376. DOI: 10.1016/j.carbon.2009.09.038

[47] Li D, Pan L, Qian J, Liu D. Highly efficient synthesis of carbon nanocoils by catalyst particles prepared by a sol-gel method. Carbon. 2010;**48**:170-175. DOI: 10.1016/j.carbon.2009.08.045

[48] Wang W, Yang K, Gaillard J, Bandaru PR, Rao AM. Rational synthesis of helically coiled carbon nanowires and nanotubes through the use of tin and indium catalysts. Advanced Materials. 2008;**20**:179-182. DOI: 10.1002/adma.200701143

[49] Cervantes-Sodi F, Vilatela JJ, Jiménez-Rodríguez JA, Reyes-Gutiérrez LG, Rosas-Meléndez S, Íñiguez-Rábago A, et al. Carbon nanotube bundles self-assembled in double helix microstructures. Carbon. 2012;**50**:3688-3693. DOI: 10.1016/j.carbon.2012.03.042

[50] Waring CE, Horton WS. The kinetics of the thermal decomposition of gaseous tetramethyltin. Journal of the American Chemical Society. 1945;**67**:540-547. DOI: 10.1021/ja01220a014

[51] Dyagileva LM, Mar'in VP, Tsyganova EI, Razuvaev GA. Reactivity of the first transition row metallocenes in thermal decomposition reaction. The Journal of Organic Chemistry. 1979;**175**:63-72. DOI: 10.1016/S0022-328X(00)82299-8

[52] Suda Y, Maruyama K, Iida T, Takikawa H, Ue H, Shimizu K, et al. High-yield synthesis of helical carbon nanofibers using iron oxide fine powder as a catalyst. Crystals. 2015;**5**:47-60. DOI: 10.3390/cryst5010047

[53] Sugioka Y, Suda Y, Tanoue H, Takikawa H, Ue H, Shimizu K, et al. Effects of dielectric barrier discharge treatment conditions on the uprightness of carbon nanofibers. IEEE Transactions on Plasma Science. 2012;**40**:1794-1800. DOI: 10.1109/TPS.2012.2199769